Printed in the United States of America

First Printing, 2020

Second Edition, 2021

ISBN 978-0-9904383-6-6

surveyingmanual@gmail.com

FUNDAMENTALS OF SURVEYING: EXAM STUDY MANUAL

(2ND EDITION)

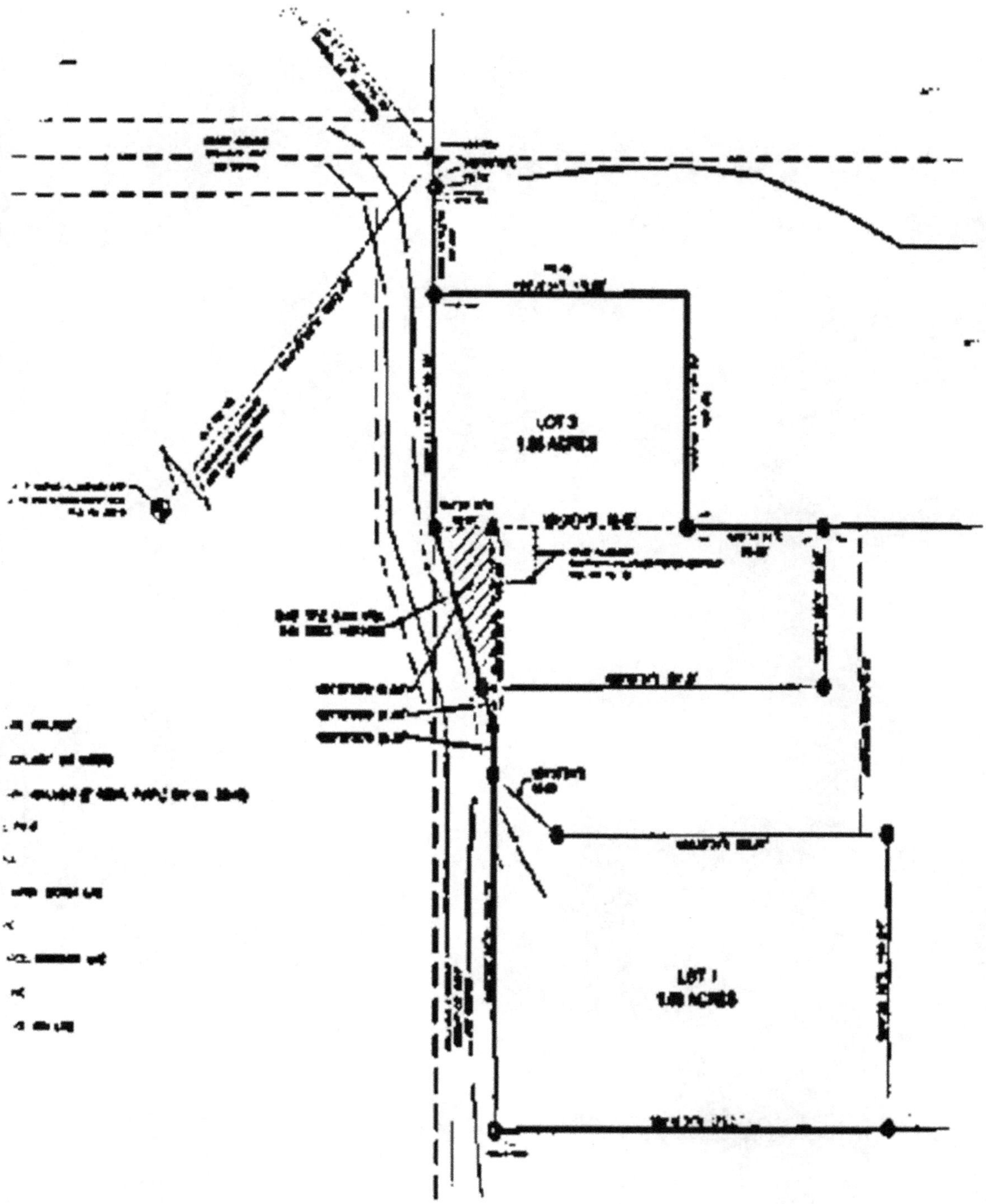

Dane M. Courville, P.L.S.

PREFACE

I began surveying in the State of Maine in 2005. On day one, I knew I was a surveyor at heart. The job and all it entails was a perfect fit for me and my life. The ability to be outside, perform historical research, perform mathematical computations, meeting landowners and their neighbors, learning to identify trees, cutting line, investigating and assisting with legal matters, and learning new equipment and technology, all seemed to be right up my alley. I graduated from the University of Louisiana at Lafayette with a degree in Biological Anthropology with a focus on forensic and biological sciences. After graduation I became a licensed elementary Montessori teacher and taught for a year in Sugar Land, Texas. Following teaching, I volunteered for a summer in Ely, Nevada and Grangeville, Idaho with the Bureau of Land Management. My volunteer efforts quickly turned into a full time job, and I stayed in central Nevada for a few years. With the Bureau of Land Management I was performing archaeological duties and working with the recreational department mapping archaeological sites and trails. I learned how to use mapping and survey equipment throughout the process. After departing Nevada for a move to Maine in 2005, I found employment with Ginn Land Surveying and Plisga and Day Land Surveyors in Bangor, Maine. With these firms I learned a tremendous amount of surveying knowledge and a skill set that put me on course to becoming licensed. I became licensed in the State of Maine in 2007 as a Land Surveyor in Training (LSIT) and finally a Professional Land Surveyor (PLS) in 2009. Leaving Maine led me back to Texas where I gained my Registered Professional Land Surveyor (RPLS) in 2010. I transferred to Colorado in 2015, and subsequently obtained my professional land surveyor licenses in Vermont, Colorado, and Oregon. In all, I have gained 16 years of survey experience working in many different survey disciplines (e.g., residential, commercial, ALTA, energy sector, transportation, construction, etc.).

Throughout my career I slowly gained the education, confidence, and wherewithal needed to focus on my roots as a teacher. The purpose of my efforts in the production and distribution of this manual is to teach the skill set needed to move forward and take the steps in becoming a licensed surveying professional. The field of surveying, across the nation, is losing ground as far a new recruits entering the profession compared to retirees leaving. Land surveying is a broad field with much to offer in regards to financial and dependability aspects. Not only is surveying diverse in its internal disciplines (e.g., math, law, history, woodsmanship, technology, art, client interaction, etc.); it is far reaching in its application (residential, governmental, energy, GIS, forensics, land development, etc.). There are too many smart, capable, and active surveyors in our grasp that are unlicensed. I am the first to admit certifications do not make you a surveyor, but certifications do open the doors for your career. A professional ceritification allows for obtaining qualifications and better opportunities for negotiating career growth and transitions. Having a professional certification only states you are “minimally competent” to performs the duties, but it is also a “ticket for never being unemployed”, as Mr. Ginn told me when I recieved my first license. This manual may not be the limits of your necessary study materials, but it will serve as a great addition. The content is outlined and organized in a format that will allow for easy information recovery and use.

Please review the material I have put together, and I wish you the very best on your journey to becoming an LSIT and eventually a PLS. Enjoy the path forward and learn from it.

If you have any questions or concerns regarding the material presented in this manual, feel free to reach out to me at surveyingmanual@gmail.com.

Thanks for your support,

TABLE OF CONTENT

While this manual is intended for study purposes with entry level land surveying experience and knowledge, it assumes a general and basic understanding of land surveying procedures, terminology, and mathematics. Suggestions for additional land surveying references can be found in the Conclusion and References section at the end of this manual.

UNITED STATES OF AMERICA:
THE PUBLIC LANDS SURVEY SYSTEM (PLSS)

Metes and Bounds States
(Territory of Thirteen Original Colonies)[11]

Connecticut
Delaware
Georgia [3]
Kentucky
Maine
Maryland
Massachusetts
New Hampshire
New Jersey
New York
North Carolina
Pennsylvania
Rhode Island
South Carolina
Tennessee
Texas [8]
Vermont
Virginia
West Virginia

Public Lands States

Alabama[1]
Alaska
Arizona
Arkansas
California [2]
Colorado [7]
Florida
Hawaii [4]
Idaho [9]
Illinois
Indiana
Iowa
Kansas
Louisiana [5]
Michigan [6]
Minnesota
Mississippi
Missouri
Montana
Nebraska
Nevada
New Mexico [7]
North Dakota
Ohio [8]
Oklahoma
Oregon [9]
South Dakota
Utah
Washington [9]
Wisconsin [10]
Wyoming [9]

1- Adopted Spanish land claims along coastal regions prior to PLSS
2- Adopted Ranchos (Spanish and Mexican Land Grants) prior to PLSS
3- Central and Western portions were surveyed into grids of Land Lots
4- Adopted the Kingdom of Hawaii native system prior to PLSS
5- Adopted French and Spanish Arpents prior to PLSS
6- Adopted French Long Lots prior to PLSS
7- Adopted Spanish Land Grants prior to PLSS
8- Adopted Metes and Bounds methods in areas
9- Adopted Donation Land Claims
10- Adopted French settlements and surveys in areas
11- Non-Public Lands are considered "Colonial" and are surveyed utilizing methods established by the original thirteen colonies that settled the new lands. These methods varied per colony due to the settlers being from different lands and cultures of Europe, but are all called "Metes and Bounds". METES- measurements along the boundary and BOUNDS- name of designation of ownerships along said line. Example of METES- Thence North 32 degrees West for 201 feet. Example of BOUNDS- Along south line of Smith Tract.

PUBLIC LAND SURVEY SYSTEM: A BRIEF HISTORY

May 20, 1775: The Continental Congress passed an ordinance calling for a method of locating and resolving land boundaries in the Western US to be established.

1785: The method of survey known as the Rectangular Survey System was created per the Land Ordinance of 1785. Following the end of the American Revolution and the subsequent Treaty of Paris in 1783; the newly formed nation of the United States had uncharted territory needing to be located, mapped, inventoried, and ultimately sold off to the citizens.

1787: The Northwest Ordinance was established by the Continental Congress.

May 18, 1796: Congress provided for the appointment of the Surveyor General.

April 25, 1812: Land surveying affairs were appointed under the jurisdiction of the General Land Office (GLO). Said GLO was initially a function of the Department of the Treasury, and then later it was designated a function of the Department of the Interior.

The Act of July 16, 1946: Abolished the GLO and reassigned all surveying duties to the newly formed Bureau of Land Management (BLM).

Survey Manual Editions:

- Manual of 1890: *Manual of Surveying Instructions for the Survey of the Public Lands of the United States and Private Land Claims*; prepared by the Commissioner of the General Land Office.

- Manual of 1902: *Manual of Surveying Instructions for the Survey of the Public Lands of the United States and Private Land Claims*; prepared by the Commissioner of the General Land Office.

- Manual of 1930: *Manual of Surveying Instructions for the Survey of the Public Lands of the United States 1930*; prepared by the Commissioner of the General Land Office.

- Manual of 1947: *Manual of Surveying Instruction 1947*; prepared by the Department of the Interior Bureau of Land Management.

- Manual of 1973: *Manual of Instruction for the Survey of the Public Lands of the United States 1973*; prepared by the Bureau of Land Management.

- Manual of 2009: *Manual of Surveying Instructions for the Survey of the Public Lands of the United States 2009*; prepared by the United States Department of the Interior Bureau of Land Management Cadastral Survey.

PUBLIC LAND SURVEY SYSTEM:
PRINCIPAL MERIDIAN LIST

Principal Meridian Name	Adopted	State containing Initial Point
Black Hills Meridian	1878	South Dakota
Boise Meridian	1867	Idaho
Chickasaw Meridian	1833	Mississippi
Choctaw Meridian	1821	Mississippi
Cimarron Meridian	1881	Oklahoma
Copper River Meridian	1905	Alaska
Fairbanks Meridian	1910	Alaska
Fifth Principal Meridian	1815	Arkansas, Iowa, Minnesota, Missouri, North Dakota, & South Dakota
First Principal Meridian	1819	Ohio & Indiana
Fourth Principal Meridian	1815	Illinois
Fourth Principal Extended Meridian	1831	Minnesota & Wisconsin
Gila and Salt River Meridian	1865	Arizona
Humboldt Meridian	1853	California
Huntsville Meridian	1807	Alabama & Mississippi
Indian Meridian	1870	Oklahoma
Kateel River Meridian	1956	Alaska
Louisiana Meridian	1807	Louisiana
Michigan Meridian	1815	Michigan & Ohio
Mount Diablo Meridian	1851	California & Nevada
Navajo Meridian	1869	Arizona
New Mexico Principal Meridian	1855	Colorado & New Mexico
Montana Principal Meridian	1867	Montana
Salt Lake Meridian	1855	Utah
San Bernardino Meridian	1852	California
Second Principal Meridian	1805	Illinois & Indiana
Seward Meridian	1911	Alaska
Sixth Principal Meridian	1855	Colorado, Kansas, Nebraska, South Dakota, & Wyoming
Saint Helena Meridian	1819	Louisiana
Saint Stephens Meridian	1805	Alabama & Mississippi
Tallahassee Meridian	1824	Florida & Alabama
Third Principal Meridian	1805	Illinois
Uintah Meridian	1875	Utah
Umiat Meridian	1956	Alaska
Ute Meridian	1880	Colorado
Washington Meridian	1803	Mississippi
Willamette Meridian	1851	Oregon & Washington
Wind River Meridian	1875	Wyoming

Note: Per *Manual of Instruction for the Survey of the Public Lands of the United States 1973*

PUBLIC LAND SURVEY SYSTEM: AREAS

1 acre=	43560 square feet=	10 square chains
1 square mile=	640 acres	
1 hectare=	2.47105 acres	
¼ acre=	1 square rood	
1 arpent=	0.8507 acres (Arkansas and Missouri)	
1 arpent=	0.845 acres (Louisiana)	
1 arpent=	0.84625 acres (Mississippi, Alabama, and Florida)	
24 square mile quadrangle=	1920 square chains	
6 square mile township=	480 square chains	
1 square mile section=	80 square chains	
1/2 square mile quarter section=	40 square chains	
1/4 square mile sixteenth section=	20 square chains	

PUBLIC LAND SURVEY SYSTEM: LENGTHS

1 mile=	5280 feet=	80 chains	
1 chain=	66 feet		
1 chain=	100 links=	0.66 feet	
1 chain=	4 rods=	4 poles=	4 perches
1 rod/pole/perch=	16.5 feet		
1 vara (Florida)=	33.372 inches		
1 vara (Texas)=	33.33333 inches		
1 vara (Southwest US)=	32.99312 inches.		
1 U.S. foot=	12/39.37 meters *(though; this is legislated per state)*		
1 international foot=	0.3048 meters *(though; this is legislated per state)*		

QUADRANGLES AND TOWNSHIPS: STANDARD PARALLELS AND GUIDE MERIDIANS; THE 4 TOWNSHIPS X 4 TOWNSHIPS/24 SQUARE MILE QUADRANGLES

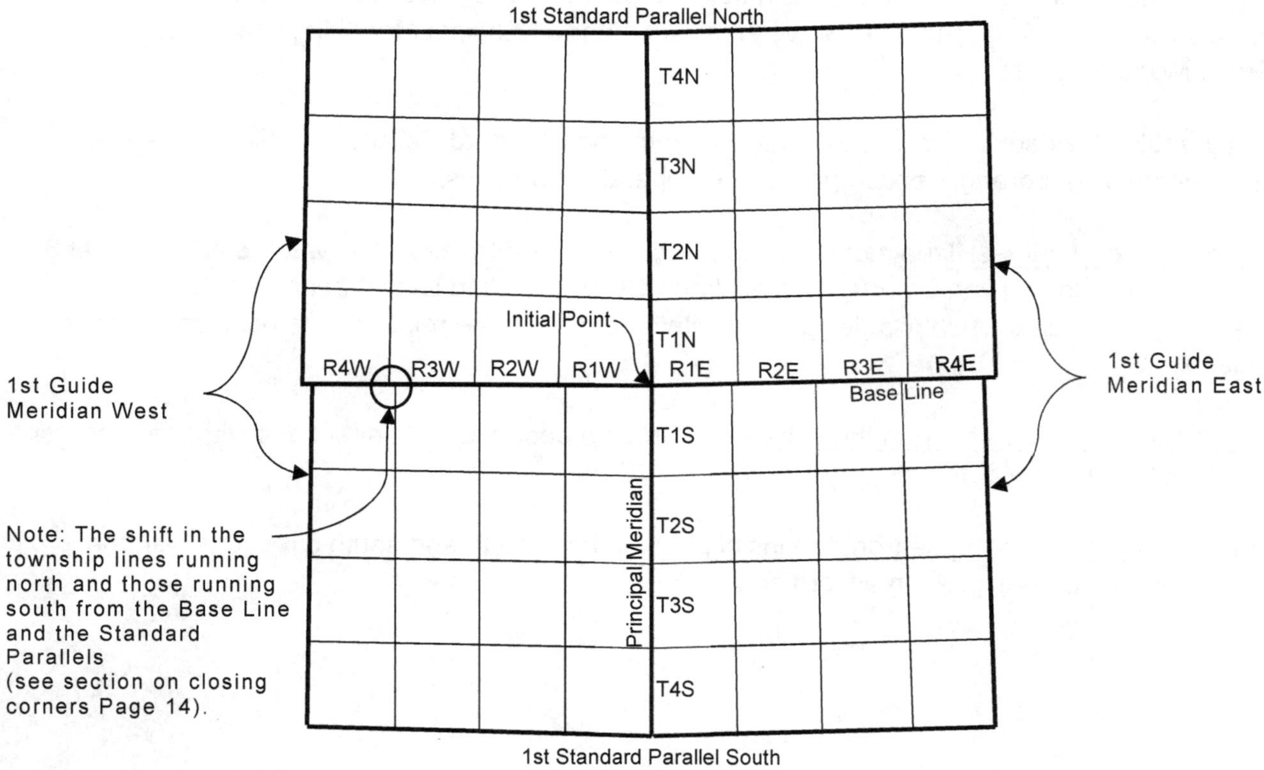

<u>Principal Meridian</u>: Primary North and South direction line of a region (intended to be along the true longitudinal meridian).

<u>Base Line</u>: Primary East and West direction line of a region (intended to be along the true latitudinal meridian).

<u>Initial Point</u>: The intersection of the Principal Meridian and Base Line. This is the origin for a region's grid system.

<u>Standard Parallels/Correction Lines</u>: East and West direction lines are parallel with the Base Line and are set at 4 township or 24 mile intervals North and South of the Base Line. They are assumed to be along the true meridian. Standard Parallels do not terminate at the intersection of Guide Meridians; they are continuous. The Standard Parallel designation is determined by the number of intervals from the Base Line (e.g., the third Standard Parallel North of the Base Line is called the "Third Standard Parallel North").

Guide Meridians: North and South direction lines being parallel with the Principal Meridian and are set at 4 township or 24 mile intervals East and West of the Principal Meridian. They are assumed to be along the true meridian. Guide Meridian lines terminate at the intersection of Standard Parallels and restart with a shift to account for the correction in Earth's longitudinal line convergence. The Guide Meridian designation is determined by the number of intervals from the Principal Meridian (e.g., the third Guide Line West of the Principal Meridian is called the "Third Guide Meridian West").

Quadrangle: A 24 square mile area established by the Standard Parallel and Guide Meridian grid system. A quadrangle encompasses 16 separate townships.

Township exterior lines: True meridian lines (e.g., north, south, east, and west) established at 6 mile intervals along given Standard Parallel and Guide Meridian lines that are establishing the quadrangles. Each township is designated with a specific name related to its position from the initial point.

Range numbers increase from the initial point: In both east and west directions with the "east" or "west" designation accordingly.

Township numbers increase from the initial point: In both north and south directions with the "North" or "South" designation accordingly.

ORDER OF TOWNSHIP LINES ESTABLISHMENT:
AFTER STANDARD PARALLELS AND GUIDE MERIDIANS ARE RUN

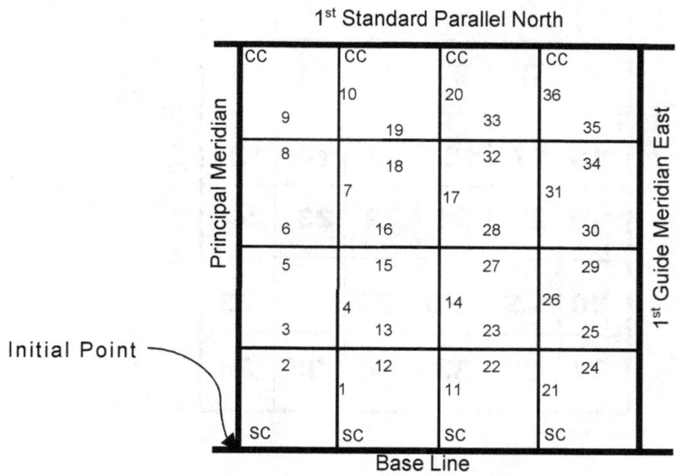

The illustration above shows the order in which the lines were originally run by the surveyor(s) establishing the township perimeter boundaries of a given quadrangle. The movement of the survey starts in the southwest corner of a quadrangle and moves towards the northeast corner then repeats south to north in a "columned" movement easterly and finishes in the northeast corner.

Note: The identification of the Standard Corner SC locations and the Closing Corner CC locations in the exhibit above. Refer to the section below entitled Closing Corners for the explanation of the establishment and information associated with Closing Corners versus Standard Corners.

SECTIONS:
TYPICAL REGULAR TOWNSHIP ARRANGEMENT

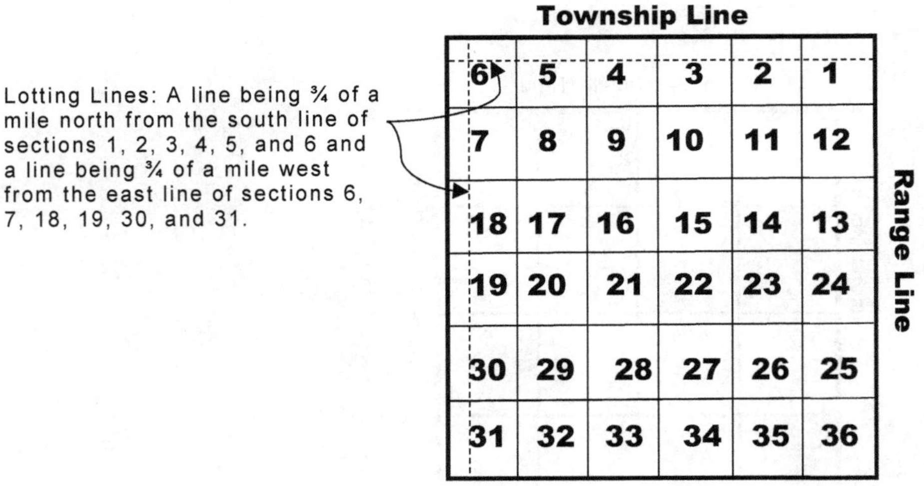

<u>Section Lines</u>: Within each 6 square mile township, section lines are established at 1 mile intervals in both north/south and east/west directions from the south and east exterior township boundaries. See the exhibit and section "Order of Section Line Establishment" for the order of section line establishment.

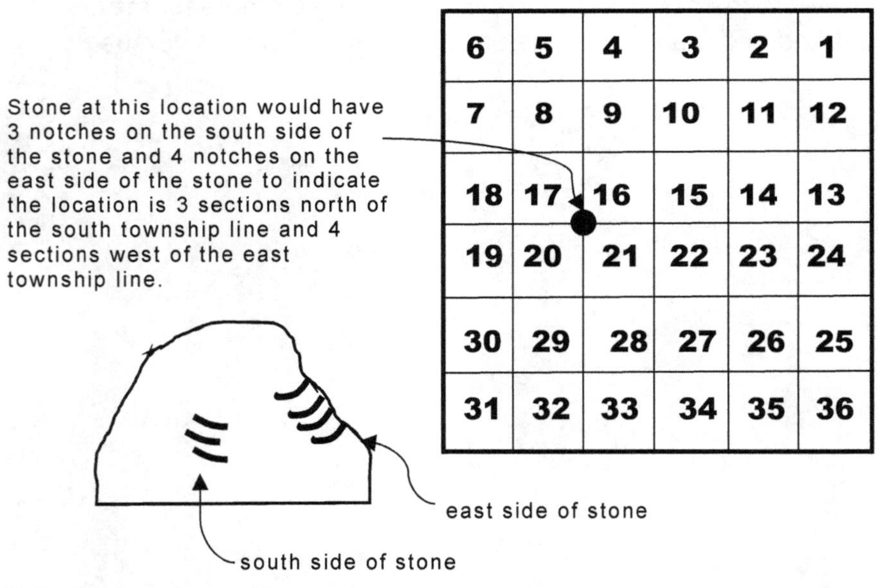

There were stones set at the intersections of sections (section corners). These stones were marked with notches to indicate the distance to the south and east boundaries of a given township.

ORDER OF SECTION LINES ESTABLISHMENT:
AFTER OUTER TOWNSHIP LINES ARE RUN

The sequence of line establishment throughout a typical township

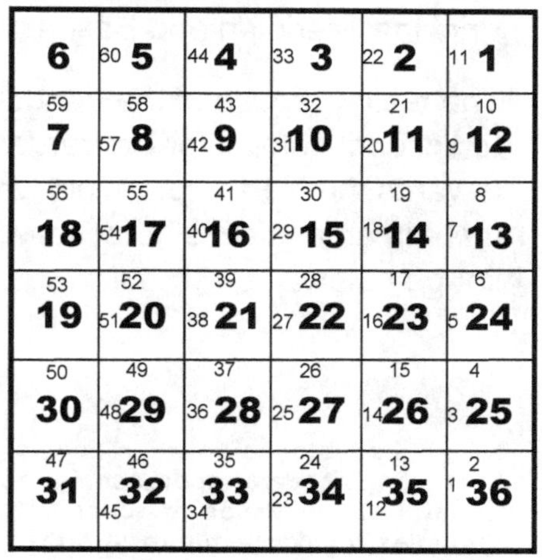

<u>The process of section line establishment is performed through the Random and True method and is as follows:</u>

<u>Random Line</u>: A "trial" line ran between two known and previously established monuments. The purpose was to obtain latitude and departure differences between the monuments. Once the information was gathered, the line is reestablished. While running the Random Line/"trial" line temporary mid-point corners are set; after the departures and latitudes of the monuments are reviewed then these temporary corners are shifted to the newly calculated locations.

<u>True Line</u>: A line ran back after a random line is initially surveyed. Once the information from the initial random line is processed, the True Line can then be established. The True Line will have a constant bearing and is known as a Rhumb Line.

<u>Method</u>
First: Line 1 is run northward (along a true bearing) for one mile from on the south line of the already established township boundary for a corner one mile west of the east line of the already established township boundary. A temporary corner is set at a mid-point (1/2 mile) during the line establishment and a temporary corner for the northwest corner of Section 36 is set.

Second: A random line for line 2 is run eastward (along a true bearing) in an attempt to connect with the already established corner on the east township line being one-mile north of the southeast corner of the township. A temporary corner is set at a mid-point (1/2 mile). A True Line is then run back and shifts both the temporary mid-point corner along Section 36's north line and the northwest corner of Section 36. These two lines establish Section 36; now on to Section 25 with the same procedure. So forth and so on goes the process moving in the order depicted in the exhibit above.

Note: All "slack/overages" or deficiencies of measurements are pushed along the last ¼ mile of each section along the north and west sides of the township. Sections 6, 7, 18, 19, 30, and 31 contain the fractional lots from moving east to west and Sections 1, 2, 3, 4, 5, and 6 contain the fractional lots from moving south to north. Notice Section 6 gets affected by both "mis-measurement" movements.

CLOSING CORNERS

Both Standard Parallels and Guide Meridians are run with the same "quality" and manner as the Principal Meridian and Base Line are run.

The main purpose of the quadrangles formed by the Standard Parallels and Guide Meridians is to account/correct for convergence angles. Lines of longitude converge as they extend north towards the North Pole. This would cause smaller and smaller townships to occur (squeezing them in an east/west direction) if Correction Lines/Standard Parallels were not in place (see image below):

Note: The distance between lines of longitude is greater approaching the Equator and is lesser approaching the poles (convergence toward the poles).

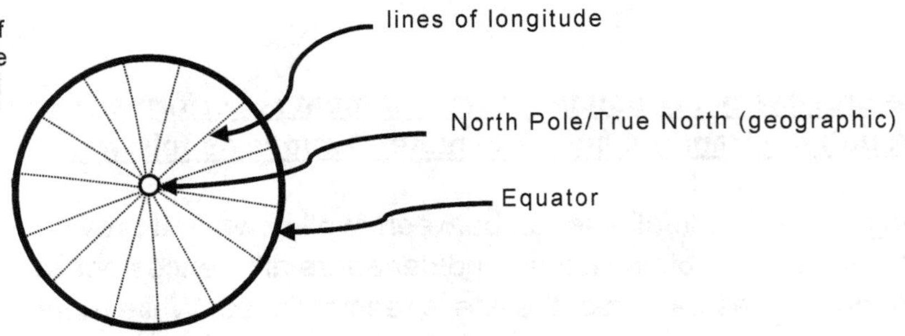

The corners established for the township corners, section corners, and quarter section corners along the north line of a quadrangle between two Guide Meridians are called closing corners. These closing corners will likely NOT align with the corners along the south line of the quadrangle to the north. There will be a shift that corrects for the convergence of the lines of longitude.

<u>Closing Corner</u>: Township corners, section corners, and quarter section corners established along the <u>north line</u> of a quadrangle.

<u>Standard Corner</u>: Township corners, section corners, and quarter section corners established along the <u>south line</u> of a quadrangle.

Note: Closing corners control DIRECTION ONLY. The true position is the point of intersection of the bearing established by the closing corner and the true Standard Parallel line (see example situation of a closing corner below).

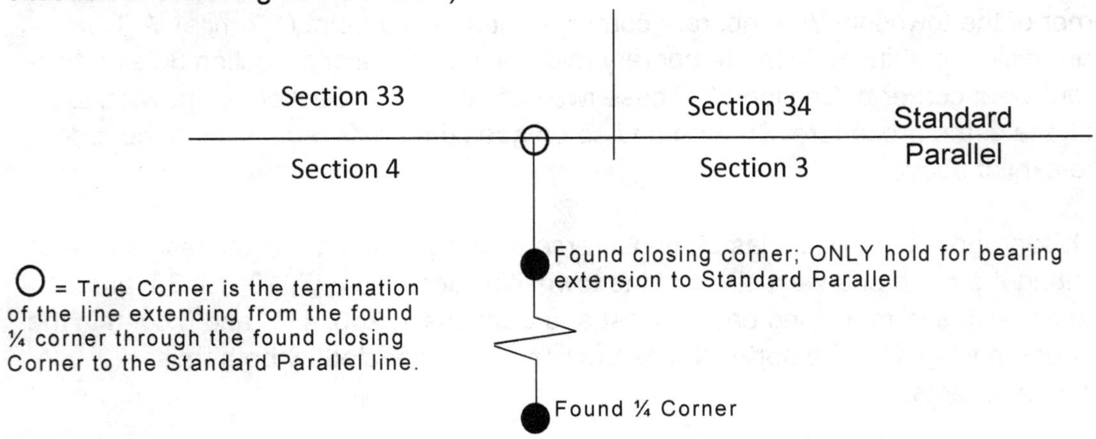

PUBLIC LAND SURVEY SYSTEM: LEGISLATED RULES

1. Once boundaries and subdivision lines are established and legally approved by authorities the lines are set and are <u>unchanging</u>. Notes and plats are constituted/approved documentation of the official boundaries and are final. Change may occur by authorities only (currently, the authority lies within the Chief of Cadastral Survey with the BLM).
2. Original and established monuments take precedence over noted bearings and distances, (<u>original monuments are paramount!</u>).
3. <u>Quarter-Quarter Section Corners</u> (if not originally established) are placed on a straight line and midway between the quarter section and section corner monuments.
4. <u>Centerlines</u> of the section are straight lines connecting the four quarter section monuments.
5. Regarding fractional sections where opposite/corresponding quarter section monuments are not available; then the centerline of the section must be run to the available quarter section monument in a cardinal direction to the opposing meander line, reservation, or other irregular boundary.
6. Whenever possible, lost or obliterated corners of approved surveys are to be restored to the original locations.

Note: Rules are assumed for "regular" sections and established with "regular"/legislatively performed survey methods.

Note: Per *Manual of Surveying Instructions for the Survey of the Public Lands of the United States 2009*; prepared by the United States Department of the Interior Bureau of Land Management Cadastral Survey.

SINGLE AND DOUBLE PROPORTIONING

Proportioning is only used when <u>all</u> other efforts for evidence recovery of a lost corner are exhausted. **PROPORTIONING IS A LAST CASE RESORT.**

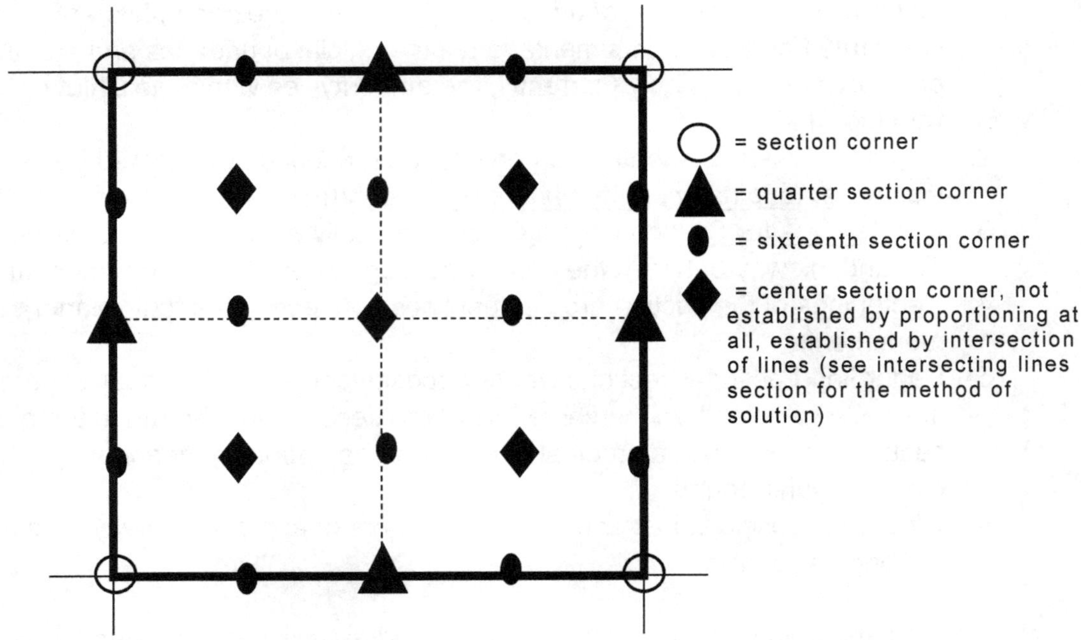

<u>When to use single vs. double proportioning</u>

<u>Single proportioning "two-way" proportioning:</u>

- All corners along a Standard Parallel/Correction Line (section, quarter section, sixteenth section, and meander corners)
- All section and quarter section corners on township boundaries
- All quarter section corners on section boundaries
- All meander corners set on a line projected across the meanderable body
- All sixteenth corners

<u>Double proportioning</u>

- All township corners
- All section corners (but restoration control does not extend outside of a given township; <u>stay within</u> a township)

If proportioning is necessary, the following acceptable evidence can be utilized

- Township corner
- Section corner
- Quarter section corner
- Closing corner
- Witness corner
- Witness point
- Line tree
- Meander corner
- Mile post
- Donation land claim corner
- Mineral survey corner

The order of reestablishment (priority given) moves from BIG to SMALL

1. Standard parallels
2. Township exteriors
3. Section corners
4. Quarter section corners
5. Meander corners
6. Minor section subdivisions

Note: Grant and Claim corners should be considered and applied in the order of priority depending on the chronological order in which they were established in relation to the evidence noted above.

Method of single proportioning

The concept behind proportioning is comparing measured to recorded lengths. The process is to first establish a ratio between the overall measured length divided by the overall record length. The next step is to figure what portion or percentage of the overall line needs to be reestablished. The final step is to set them equal to each other and solve for the missing element. The ratio of the totals equals the ratio of the proportioned.

First: $\frac{measured\ total}{record\ total}$ Second: $\frac{measured\ portion}{record\ portion}$ Third: $\frac{measured\ total}{record\ total} = \frac{measured\ portion}{record\ portion}$

Example 1

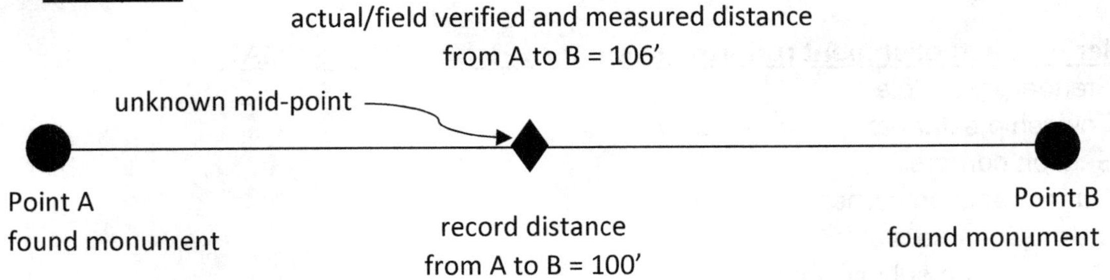

Total measured distance = 106'
Total record distance = 100'

Objective is to reestablish the mid-point between the found monument at point A and the found monument at point B.

Step 1: Set up to ratio of measured vs. recorded distances. In single proportioning this ratio needs to be established in order to prorate that percentage to the proportional length.

$$\frac{measured\ total}{record\ total}$$

$\frac{106'}{100'} = 1.06$ Note: The record is smaller than the measured so the ratio is larger than one.

Step 2: Establish the percentage/portion of the overall line that needs to be determined. In this instance, the mid-point is needed; meaning half or 50% of the line.

$$\frac{measured\ portion}{record\ portion}$$

$\frac{x}{50}$ Note: The measured length is unknown, but the record should be 50', because 50' is half of 100'.

Step 3: Set the two proportion ratios (overall and portion) equal to each other and solve for the unknown element.

$$\frac{measured\ total}{record\ total} = \frac{measured\ portion}{record\ portion}$$

$$\frac{106'}{100'} = \frac{x}{50'}$$

$$1.06 = \frac{x}{50'}$$

$$53' = x$$

The measured distance half way between the two found monuments is 53'. The distance between the monuments should have been 100' per the record, but when measured the distance was 6' longer. That said the mid-point should have been 50', but given the 6' "bust" the true mid-point is 53'.

Example 2

In the example, find the location of a monument to be set a quarter of the way between A and B (from point A).

$$\frac{measured\ total}{record\ total} = \frac{measured\ portion}{record\ portion}$$

$$\frac{106}{100} = \frac{x}{25}$$

$$1.06 = \frac{x}{25}$$

$$26.5 = x$$

Note: The difference between the measured and the record divided by the percentage of the length being determined plus the record portion will derive the same result.
In the example above: (6'/0.25) + 25 = 26.5'.

Method of double proportioning

The concept associated with double proportioning is to perform single proportioning TWICE.

When performing double proportioning in regards to sectionalized land, the use of cardinal equivalents is paramount.

Cardinal equivalents: Distances along TRUE bearings. Due north, south, east, or west.

The change in y values along cardinal (north/south) is known as LATITUDE
The change in x values along cardinal (east/west) is known as DEPARTURE

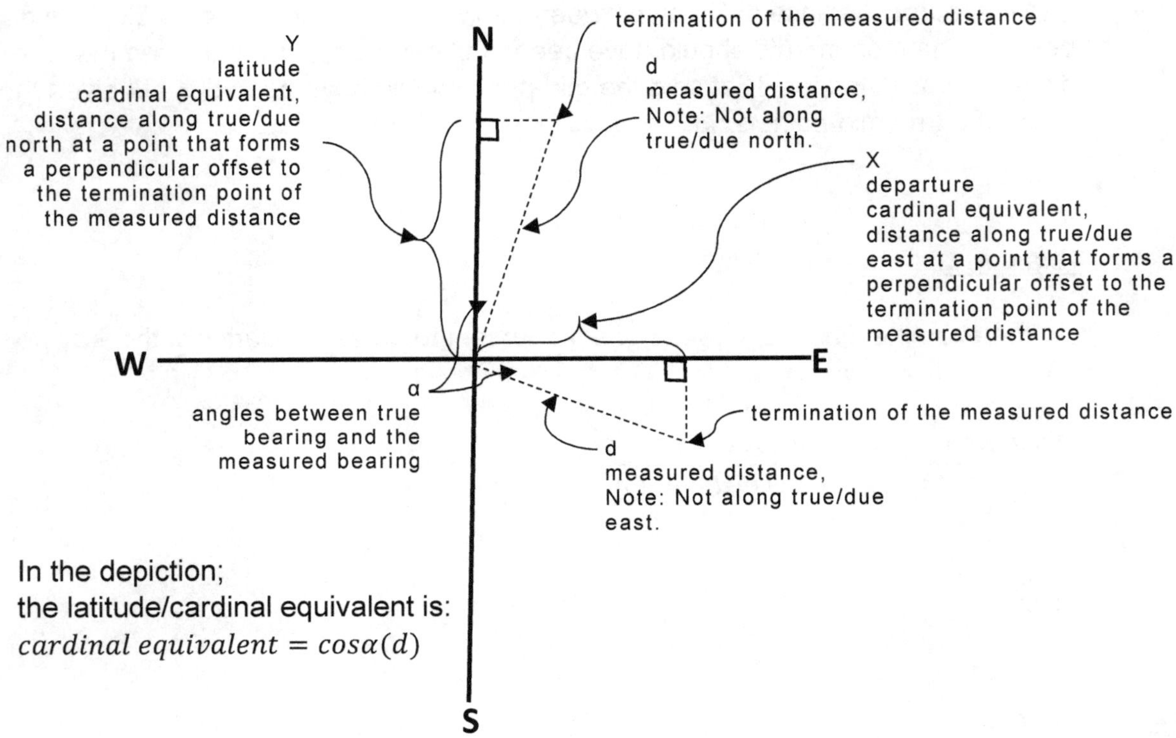

In the depiction;
the latitude/cardinal equivalent is:
$cardinal\ equivalent = \cos\alpha(d)$

To solve for cardinal equivalent values simply utilize right triangle solutions. The determination of a cardinal equivalent is found by knowing the measured distance (hypotenuse), the angle between true and measured bearings, and finally solving for the adjacent or opposite legs of the right triangle that is formed accordingly.

Once cardinal equivalent values are determined from the acquired or provided measured distances, then a single proportioning can be performed for both the north/south direction and then the east/west direction. The location of these two single proportioning positions will indicate the lost corner location.

Procedure

Determine departure and latitude values.

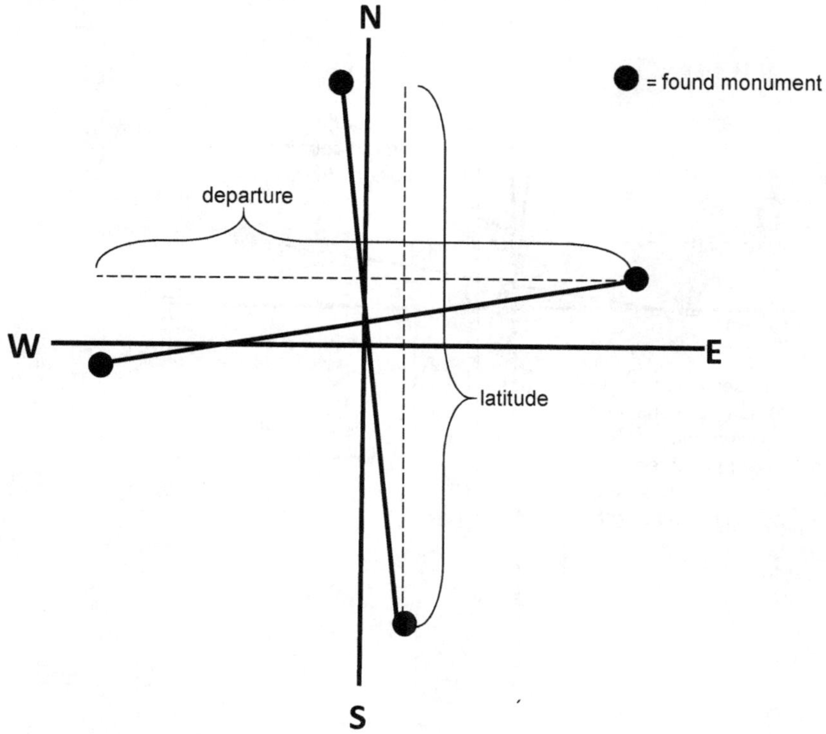

Determine the single proportion value/location along both the departure and the latitude.

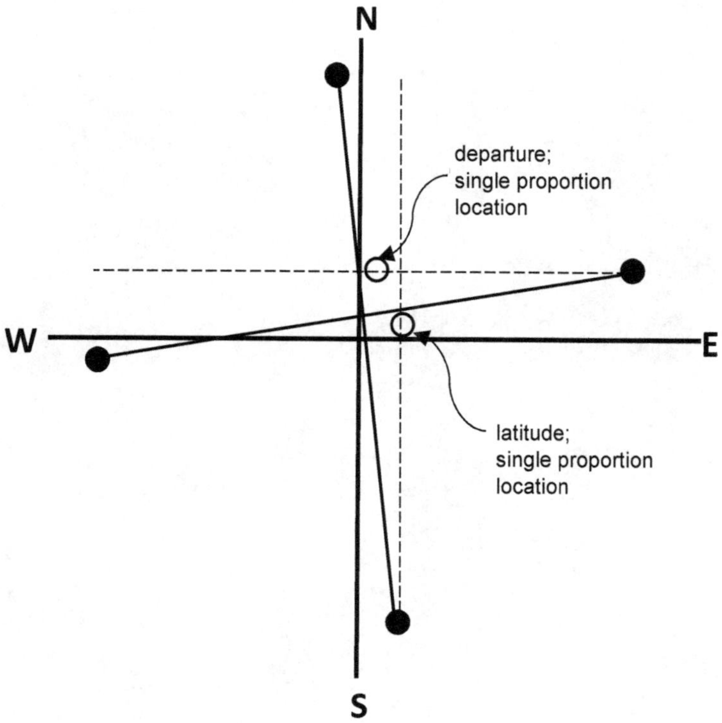

Determine intersection of the single proportion valve/locations.

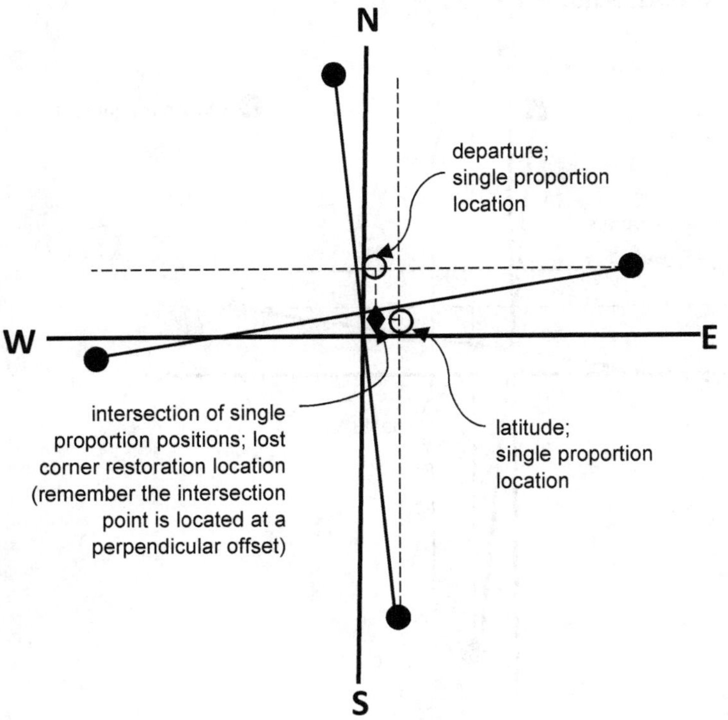

Example

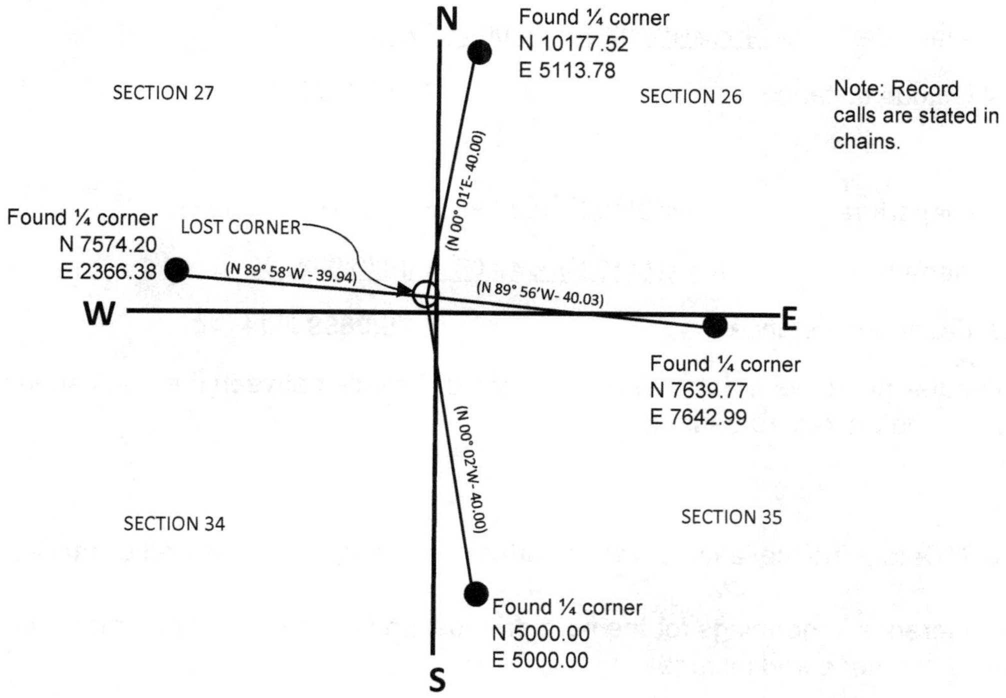

Step 1: Determine latitude and departures for the record calls.

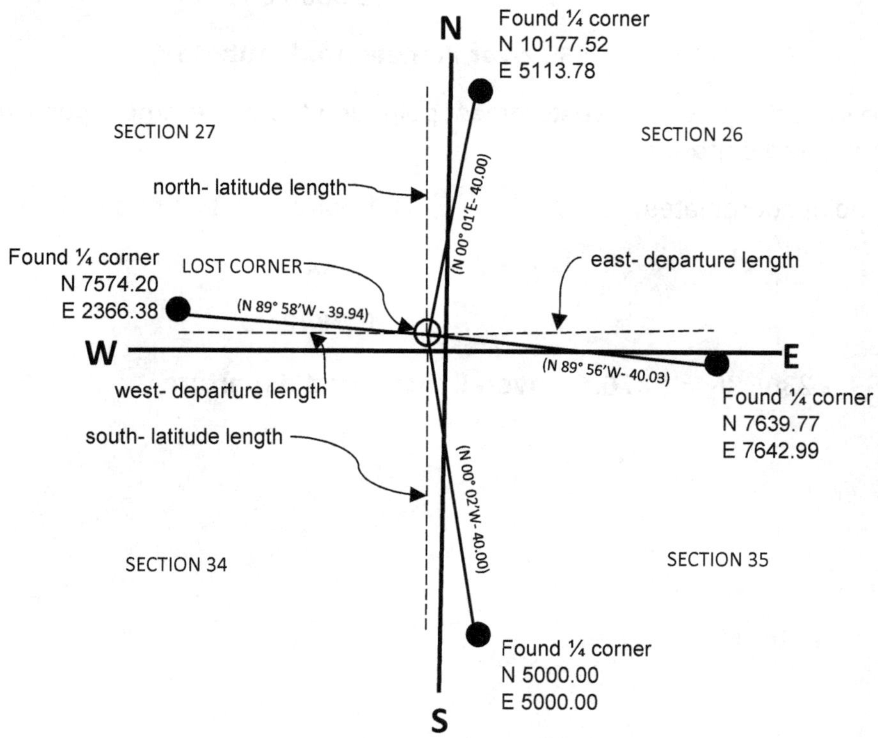

North- latitude $\cos(00°01') \times 40.00 = 39.99999831$

South- latitude $\underline{+\cos(00°02') \times 40.00 = 39.99999323}$

Total latitude distance **79.99999154**

West- departure $\cos(00°02') \times 39.94 = 39.93999324$

East- departure $\underline{+\cos(00°04') \times 40.03 = 40.02997290}$

Total departure distance **79.96996614**

The angles noted were determined to be the difference between the true/cardinal bearing and the record bearing.

Step 2: Determine the latitude and departure values for the measured distances.

The difference in northings for the northernmost point and the southernmost point will provide the measured latitude.

Northernmost point coordinates
N **10177.52**

E 5113.78

Southernmost point coordinates
N **5000.00**

E 5000.00

$10177.52 - 5000.00 = \mathbf{5177.52}$ **overall measured latitude**

The difference in eastings for the westernmost point and the easternmost point will provide the measured departure.

Westernmost point coordinates

N 7574.20

E **2366.38**

Easternmost point coordinates

N 7639.77

E **7642.99**

$7642.99 - 2366.38 = \mathbf{5276.61}$ **overall measured departure**

Step 3: Determine the single proportioned location along both the north/south line and the east/west line. To perform this step use either of the two record departure lines to determine the departure single proportion location and use either of the two record latitude lines to determine the latitude single proportion location.

Latitude

$$\frac{measured\ total\ latitude}{record\ total\ latitude} = \frac{measured\ portion\ latitude}{record\ portion\ latitude}$$

$$\frac{5177.52}{79.99999154} = \frac{measured\ portion\ latitude}{39.99999323\ (selected\ south\ line)}$$

$$\frac{5177.52}{79.99999154} \times \frac{39.99999323}{1} = 2588.760$$

Now, because the south line was used to proportion from (could also have been north line) we need to add 2588.76 to the northing of 5000.

$2588.760 + 5000.00 = \mathbf{7588.76}$ **northing value of lost corner restoration position**

Departure

$$\frac{measured\ total\ departure}{record\ total\ departure} = \frac{measured\ portion\ departure}{record\ portion\ departure}$$

$$\frac{5276.61}{79.96996614} = \frac{measured\ portion\ departure}{39.93999324\ (selected\ west\ line)}$$

$$\frac{5276.61}{79.96996614} \times \frac{39.93999324}{1} = 2635.336$$

Now, because the west line was used to proportion from (could also have been east line) we need to add 2635.336 to the easting of 2366.38.

$2366.38 + 2635.336 = \mathbf{5001.72}$ **easting value of lost corner restoration position**

Finally, the coordinate for the lost corner is

N 7588.76

E 5001.72

MEANDER LINES

1) ACREAGE, ACREAGE, ACREAGE: A meander line's purpose is to assist in area calculations (are not boundaries, but merely acreage delineations).

2) Meander lines are surveyed along Ordinary High Water Marks (OHWM) on the banks of waterbodies.

3) What is NOT meandered:
    a. National parks
    b. Indian reservations
    c. Mineral claims

4) What is meandered:
    a. Large streams (3 chains in width measured perpendicularly)
    b. Navigable rivers
    c. Lakes (50 acres and upwards; used to be 25 until 1973)
    d. Ponds (50 acres and upwards; used to be 25 until 1973)

5) Once set and surveyed a meander line remains in place REGARDLESS if ordinary high water line changes over time.

EASEMENTS

According to *Brown's Boundary Control and Legal Principles* and easement is defined as- "An Interest in land created by grant or agreement that confers a right upon owners to some profit, benefit, dominion, or lawful use of or over the estate of another; it is distinct from ownership of soil."

Easements: <u>Rights</u> are granted **NOT** <u>Title.</u>

Easement documentation should state LOCATION and PURPOSE

Example: "Over the lands of Smith for horse cart transportation".

<u>Characteristics</u>

1. An interest in land that must be created by grant or agreement, expressed or implied.
2. Interest must be on the land of another (cannot have an easement over one's own land).
3. Easement or nonpossessory; easement holder can only prevent interference with their interest.
4. Privilege to use an easement must be capable of creation.
5. Must be described.

<u>Types of easements</u>

Appurtenant: Attached to LAND(S):

Dominant: Land provided an easement (benefited from the easement).

Servient: Land affected by an easement (burdened from the easement).

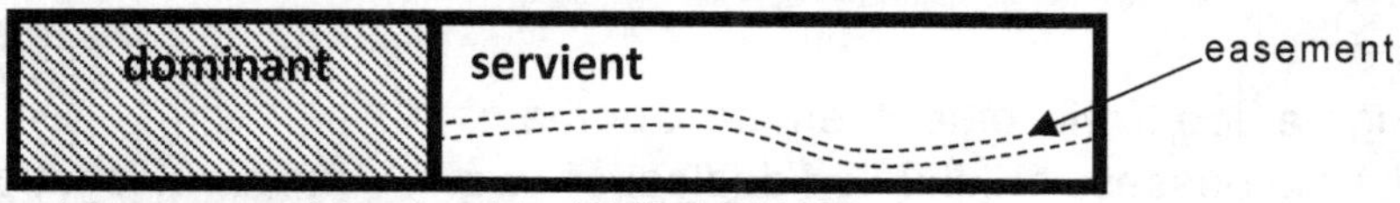

Easement in gross: Attached to PERSON(S)

<u>Who</u>

Public: ALL can use.

Private: SELECTED individuals can use.

<u>Negative or Positive (affirmative)</u>

Positive: Physical act on, under, or over land of another (e.g., road, well, crop, etc.).

Negative: Prevention of action or use of land of another (e.g., light, air, scenic overlook, etc.).

<u>Easement description types</u>

Description of dominant parcel: Describes the land it is going to.

Description of servient parcel: Describes the land it is coming from.

Description of easement itself: Describes the actual easement.

<u>Assumption</u>

It is assumed that title is to the center of an easement unless explicitly stated otherwise.

TITLE

Grantor vs. Grantee

Grantor=seller and Grantee=buyer

Real property interests

Possessory

Fee simple absolute: No encumbrances or stipulations.
Fee simple defeasible: Voided on certain expressed conditions.
Fee simple conditional: Grantee must produce heir.
Fee tail: May only pass to grantee's heir(s).
Estate of life: Use during grantee's natural life only.

Non-Possessory

Easement: Right to use another's land.
Covenant: A promise concerning land.
License: A privilege to use another's land.
Profit a pendre: Right to remove material.
Emblements: Right to use annual crop.

Encumbrances

Mortgage: Debt.
Tax lien: Non-payment of taxes.
Mechanics lien: Non-payment of services.
Statutory lien: Lien placed by statute(s).

Methods of transfer

Written: Through a deed, will, patent, eminent domain.
Unwritten: Adverse possession, implied dedication, waiver, practical location, unwritten agreement, estoppel, acquiescence, nature. Unwritten rights become senior to written rights.

Unwritten methods of transfer

Adverse possession

Elements

1. Actual possession: Record owner and adverse possessor cannot simultaneously "possess" the land.
2. Statutory period must be met: Varying by state.
3. Open/Notorious: Possession is noticeable to the record owner (not hidden but visible) and illustrates the extent of possession.
4. Exclusive: Possessor treats the land as though they are the true owner; no other individual is possessing simultaneously.
5. Hostile: Possessor takes action to prevent record owner from access (e.g., fence, gate, signage, etc.).
6. Claim of title: Possessor has a good faith claim like color of title (color of title is when a defective deed appears to provide title to possessor, but in actuality, it does not).

Burdens of adverse possession claim

1. The burden to prove adverse possession is on the possessor and not the record owner.
2. Evidence must be clear and convincing.
3. Possessor must rely on the strength of their possession as opposed to the weakness of the record owner's title.
4. Defendant (record owner) has to prove the possessor had permission.

Defenses of adverse possession claim

1. Refute any element (all it takes is one).
2. Statutory time period if extended.
   a. For government ownership (can never adverse possess against government).
   b. For insanity claims (time is extended if record owner is not mentally sound).
   c. For infancy claims (time is extended if record owner is too young).
   d. For military service time (if serving in military, then time is extended).
3. If possessor is provided permission by record owner, then claim will not hold.

Surveyor's responsibility regarding adverse possession

1. Inform client of what adverse possession is.
2. Locate the record boundaries.
3. Locate possession boundaries.
4. Provide description of adverse possession area.
5. Gain evidence of possession elements.

<u>Implied dedication</u>

Giving a gift of fee simple title, or right of use (easement), to the public without an expressed written transfer taking place.

Elements

1. Owner has the right of title to offer.
2. The offer is conducted by the actual owner.
3. The public entity must officially accept the offer made.
4. The boundaries must be defined/described.

<u>Waiver</u>

To intentionally abandon or relinquish what would ordinarily be the possessor's to claim.

Practical location

True location is not known, but the location is "believed" to be known.

A common grantor or two parties agree and monument a common boundary; any inaccuracies in the correctness of said line is to be considered a "done deal". There are no further disputes (unlike in an agreement).

Elements

1. Original boundary is vague and truly unknown.
2. The newly decided line must be monumented.

Unwritten agreement

True location is not known, and the location's uncertainty is revealed and accepted by all parties.

Defines and locates the written boundary between parties, said boundary without the agreement would be vague/unknown and be in dispute.

Elements

1. Original boundary is vague and truly unknown.
2. Boundary is in dispute between parties.
3. The resulting boundary needs to be agreed on by all parties.
4. Boundary is discernible (can possibly be detected).

Estoppel

The lawful protection of a landowner from another landowner's allegations or denials of a fact(s) that contradicts their own previous acts.

Elements

1. One puts forth omissions, declaration, words, actions, etc.
2. One's omissions, words, actions, etc. cause reasonable and foreseeable reliance by another.
3. Expenditures otherwise would not have been made having known the truth.
4. Expenditures cause injury or loss.

Acquiescence

Passive compliance or satisfaction. Boundary created by inaction or silence. The boundary is thought to be true and nothing is done in contradiction.

Elements

1. True boundary location is uncertain or in doubt.
2. Boundary location is made visible and is ascertainable.
3. Active and constructive notice to the adjoining landowner is made.
4. Lack of action or silence by the adjoining landowner to stop Acquiescence.

Nature

Erosion and avulsion of land can cause change in title when dealing with riparian lands (refer to section on riparian and littoral rights).

Note: Regarding the unwritten transfers of title outlined above; each state has legislated rules stating conditional specifications, requirements, limitations, and statutory timing. The information above is general in nature and attempts to cover the definitions and basic understanding of each type of unwritten transfer method.

DEEDS

A deed is EVIDENCE of title, but not actual TITLE.

Deed types

- Warranty: Grantor takes all the responsibility all the time.
- Special Warranty/Quit Claim with Covenants: Grantor takes responsibility only during the time of their period of ownership.
- Quit Claim/Release: Grantor takes no responsibility; statement of "I will give you everything that I was given".
- Sherriff: Like a quit claim from a county; after the statutory time period the title is guaranteed.

Deed formalities

- Must be in writing: Statute of frauds.
- Must be signed by grantor: The grantee signature is "optional".
- Acknowledgement: If to be used for record or authorization.
- Seals/certifications are not necessary for conveyance.
- Signature of spouse (if spouse has interest).
- Statement of consideration (how much/amount agreed on).
- Identification of grantee.
- Proper form/parts are required.

Parts of a deed

- Granting clause/words of conveyance- describe what is to be conveyed and to whom.
- Identification of parties.
- Consideration/amount.
- Operative wording- words that effectively transfer the interest outlined in the deed.
- Description.
- Noted exceptions and reservations.
- Covenants, acknowledgement, and signature of grantor.

DESCRIPTIONS

Description types

- Metes and Bounds: (*N 22 degrees 14 minutes for 223.56 feet)*
- Subdivision or Referral: (*Being Lot 2 of Block 1 in the Sampson Subdivision)* or (*Same lands described in Book 458, Page 623)*
- Bounds or Adjoiner: (*By and along Bear Creek)* or (*By and along Fredrickson)*
- Proportional: (*West half of Gaylord land)* or (*N/2 of the SE/4 of Section 4, T5S, R10E; 6th PM)*
- Reference to line of strip: (*10 feet east and west of the herein described centerline)*
- Area: (*Being 250 acres in the S/2 of Gael lands)*
- Monuments: (*Billy Bob land with the set rebar)*
- Exception: (*All Henderson lands except the W/2)*
- Linear: (*250 feet from the West line of Drake property)*

Descriptions are sufficient if

They clearly describe and identify land for reasonable retracement by a competent surveyor.

Qualifying vs Augmenting clauses

Qualifying: Clause used for easement, exceptions, or other rights to be REMOVED from the property.

Augmentative: Clause used for easement, exceptions, or other rights to be ADDED to the property.

EVIDENCE

Regarding boundary dispute court cases

The plaintiff has the burden of proof.
The defendant has burden of coming forward once plaintiff has met the burden of proof.

With criminal cases the burden is "beyond reasonable doubt"
(scales of justice are tipped ALL the way; must be a 100% to 0% favor)

BUT

With boundary cases the burden is "preponderance of evidence"
(scales of justice are tipped slightly; has to be at least a 51% to 49% favor)

Evidence as it pertains to records and documents

- A deed is evidence of title and not proof of title. The deed can serve in the assistance to establish title only.
- That which is clear cannot be made unclear by the addition of uncalled for evidence. If the record speaks for itself, then additional evidence should not and cannot be introduced. This is also known as the "Four Corners Rule": The document/deed should have everything needed to make a determination; one should not have to look any further than the "four corners" of the document/deed.
- Parol (verbal) evidence can never dispute written evidence. Written evidence takes precedence over parol evidence.
- The document/deed should be reviewed in its entirety and not "pieced out". The document/deed as a whole should be considered as it relates to its complete self. One cannot excerpt out parts without consideration of all the parts.
- The description should identify the property and not "define" the property. Location, location, location; not characteristics.
- External documents referenced inside a subject deed/document hold just as much weight as the subject deed itself. Simple reference to or mention of other documents within a subject deed allows for the use of said referenced documents.
- Extrinsic evidence can only be used to clarify that which is unclear. If the deed/document is clear and the intent is outlined, then extrinsic evidence cannot be used.

Types of evidence

- Parol: Verbal/Oral testimony can only be used to clarify an ambiguity. If intent is known, then parol evidence may not be used. Typically provided by a third party that will not be affected by the result of the testimony.
- Hearsay: A statement provided in court that was initially made outside of the court.
- Common report: That which is known and accepted by several individuals (common knowledge).

Regarding the above evidence, the following must be implied:

- Can only provide evidence of title, but not title itself.
- Utilized as a "last case resort".
- Cannot contradict valid written evidence.
- Individual(s) providing the evidence are considered unbiased.

Extrinsic evidence

Extrinsic evidence is that evidence that is made available from OUTSIDE of the written document/deed.

Extrinsic evidence is used to clarify intent. It helps determine location, but it will never change location. As with parol, hearsay, and common report, extrinsic evidence must be used as a "last case resort".

Uses for extrinsic evidence may include:

- Verification of monumentation or location.
- Clarification of errors, omissions, and conflict.
- Definition or term identification.
- Validate or prove documents/deeds are lost.
- Clarification of ambiguities.

Types of ambiguities:

Latent: Less obvious; not on the face of the deed (i.e., the ambiguity is revealed when the deed wording is compared to the real situation in the field).

Patent: Obvious; on the face of the deed (example: *North 33 degrees 25 seconds* ***North****, a distance of 223.65 feet*; obviously incorrect).

What can be used as extrinsic evidence

- Parol evidence.
- Photos.
- Unrecorded documents/papers.
- Contemporaneous or subsequent acts.
- Declarations of knowledge.

APPORTIONMENTS:
SIMULTANEOUS CONVEYANCES

Things to remember when apportioning

- It is used as a "last case resort"; evidence is paramount.
- Never apportion with the use of known erroneous measurements.
- Apportioning should never be extended/carried beyond found monuments.
- Never apportion road width; road widths are set at stated/recorded measurement. (This is a very common question during examination, and will catch you if you do not watch out!).
- Only apportion simultaneous conveyances (subdivided lands).

The process of apportionment is the same process as outlined in the single proportioning section. The process is the same, just the naming is different. Typically, apportionment is associated with subdivisions and proportioning is associated with sectionalized land.

RECORD RESEARCH

It is the responsibility of the SURVEYOR to perform due diligence in an effort to recover all pertinent record research in relation to the LOCATION of the subject parcel.

It is the responsibility of the LAWYER to perform due diligence in an effort to recover all pertinent record research in relation to the TITLE of the subject parcel.

When has the "right" amount of research been performed

The proper depth of research needed is determined by whether or not the boundaries have been described at the time in which they were created.

What is the purpose of record research

- Recover deed/documents in relation to location of title rights.
- Recover deed/documents in relation to the extent of title rights.
- Recover deed/documents in senior and junior right relationships.
- Discover any conflicts in location, extent, or senior/junior rights.

What is the scope of research

- Subject's records: Obtained from county/state registries (states differ in recording procedures).
- Adjoiner's records: Obtained from county/state registries (states differ in recording procedures).
- Easements of roads bounding the subject parcel.
- Private records of client, adjoiner, local surveyors, etc.

What to note or document during research

- Dates.
- Grantor.
- Grantee.
- Instrument/deed/document type.
- Instrument/deed/document identification/number/book and page.
- Special conditions.

Recording statutes of race, notice, and race/notice

- Race: First to record (with or without having notification at the time of the purchase) is the titleholder.
- Notice: Last to purchase (without having notice of previous sales) is the titleholder.
- Race/Notice: First to record (without having notification at the time of the purchase) is the titleholder.

Senior vs. junior rights

The general rule is the parcel that was created EARLIER has senior (superior) rights and LATER created parcels have junior (subordinate) rights.

A parcel of land created first chronologically has senior rights over a parcel of land that is created at a later date and has junior rights.

Having senior rights entitles a parcel of land to receive all lands described at the time of its creation. The idea is allowing a parcel to receive the land it was deeded.

Senior and Junior rights apply to sequential conveyances only and do not apply to simultaneous conveyances (i.e., subdivisions).

Example

- A 1,000,000 square foot (22.96 acre) tract currently exists. The tract is a square 1000 foot parcel.
- From this parcel, the SE/4 is separated and conveyed. It is a square 500 foot parcel (250,000 square feet; 5.74 acres).
- The newly created SE/4, 5.74 acre parcel has senior rights over the remainder 17.23 acres.
- The subject 22.96 acres has junior rights and the new 5.74 acres has senior rights if any conflicts arise.

Conflicting record/document information

That which is most certain will control over that which is derived.

Order of precedence of documentation

1. Surveyor's work and found evidence.
2. Field notes.
3. Plat/plan.
4. Prior description.
5. Current description.

Conflict within a single document

- The document stands alone ("the four corners" rule).
- Parts of the document must be harmonized.
- Consideration of when/time the document was created.
- Subsequent actions after the document's creation are taken into account.
- When in conflict, wording interpretation will favor the grantee and not the grantor (it was the grantor's responsibility to get it right in the first place).

Conflict among documents

- Handwritten over printed.
- Senior over junior.
- Recorded over unrecorded.
- Field notes over plan over deed description.
- Preponderance of evidence consideration.
- Chronological order of recordation.

RULES OF CONSTRUCTION

Rules of construction are applied if and only if intent is unclear. If intent is known, clear, and understood then the application of rules of construction is unnecessary.

Obvious: Do NOT apply rules of construction.
Unobvious: Apply rules of construction.

Order of precedence to consider

1) Intent.
2) Original location.
3) Monuments (called for in operative record). Note: Operative record is that which creates the line; the monuments represent the original line.
4) Measurements.
5) Area.

Intent

Intent is the true meaning as derived from the wording of the document/deed. Intent requires no "interpretation"; the meaning should be obvious and has no ambiguity. If intent is known and understood clearly then GO NO FURTHER with the rules of construction, there is no need to construct anything.

In the writing of a deed, it is the responsibility of the GRANTOR to lay out intent. If ambiguity does arise, then favor will be given to the grantee in any interpretations.

Original location

Original location is the location of the boundary at the time of its creation. All the evidence, monuments, etc. are simply there to help determine and indicate this original location. The evidence does not hold, the line holds.

There is a difference between a corner and a monument. The monument physically indicates the actual corner; it is not the corner. The monument marks the corner.

Monuments

Types of monumentation

Material

- Natural: Hold the highest priority (e.g., tree).
- Artificial: Hold the second highest priority (e.g., 5/8" iron rebar and cap).
- Record: Hold the least priority (e.g., adjoiner call- "*Smith east line*").

Recorded vs. Unrecorded

- Documented: Hold the highest priority; these monuments are called for in the operative record.
- Undocumented: Hold the second highest priority; these are not called for in the operative record.
- Possession: Hold the least priority (e.g., uncalled evidence of occupation, such as a fence along a property line).

Condition

- Existent: The monument is present and represents the corner, corner location is determined by original or approved replacement monument, or through accepted accessories to original or replacement monument.
- Disturbed: Monument is existent but visibly disturbed.
- Obliterated: Corner location that is not available, but can be restored through a) competent testimony, b) accepted record evidence, c) improvements built at the time the original position was known, and/or d) a monument proven to be the replacement of the original corner.
- Lost corner: Corner location that is not available and cannot be determined based on evidence. A lost corner requires mathematical analysis from other found corners for repositioning. Lost corners are the only corners requiring mathematical computation.

Rights status

- Senior: The monument is called for in earlier or original records.
- Junior: The monument is called for in later or subsequent records.

Order of precedence with monumentation evidence

1) Original monument.
2) Subsequent "chain of evidence" monument.
3) Witness monument.

Note: If monuments are not called for in the document/deed, then they simply become supplemental evidence.

Ways monumentation can represent a corner

- Locative: The monument in its typical representation to a corner (e.g., *to a pin*).
- Witness/Reference Tie: The monument holds a position from which the corner location can be derived (e.g., *to a rebar, from which the SE corner bears North 50 feet*).
- Passing: The monument holds the line between the corners (e.g., *passing a tree stump and continuing on said course to....*).

Locative vs. Descriptive

Given the following example: *Thence to a stone in a fence line.*

- The call is to the STONE and is the locative call.
- The call is descriptive in mentioning the fence line, but the FENCE LINE is descriptive; not the call.

Rules of monumentation

1) A called for monument is calling for the position the monument marks; not the monument itself. This is the corner vs. monument difference; the monument represents/marks the corner and is not the corner itself.
2) Precedence: First original, second replacement "chain of evidence", third witness.
3) Monuments are controlling only if they are:
    a. Called for in operative document.
    b. In the same position (undisturbed).
    c. Set in the condition as called for in the cited conveyance.
    d. Existent at the time of the conveyance.
4) Due to assumption of impartiality, public sector surveyor monumentation is consider higher in priority than private sector surveyor monumentation.
5) Priority
    a. Natural over artificial monuments.
    b. Called for over uncalled for monuments.
    c. Locative over passing.
        i. Witness monuments called for in writing have the same weight as an unfound corner monument.
        ii. Witness monuments help locate corner monuments and do not substitute for them.
        iii. Passing monuments hold little weight.
        iv. Witness monuments can help find corner monuments, but do not refute a found, original, and undisturbed corner monument.
    d. Documented over undocumented over possession, but undocumented and possession monuments can assist in locating a corner.
    e. Locative over descriptive.

Exceptions/Caveats to the rules of monumentation

- Disturbed monumentation loses all or some validity.
- Monuments will not hold if the grantor's intent clearly states otherwise in the written record.
- If the found monument is clearly a mistake it will not hold.
- Regarding senior title dispute, a monument will not extend a parcel's coverage beyond what it is entitled to.

Ways to identify monumentation

- Time.
- Physical characteristics.
- Markings.
- Relation to other markings.
- Public or private records.
- Parol evidence.
- Hearsay evidence.
- Common report.
- Measurements from other accepted monumentation.
- Witness or reference tie monumentation.
- Trees.

Measurements

Rules for distances and direction (measurements)

- The most certain rules.
- Measurements only assist in locating monuments.
- When in doubt, direction is assumed to be true (in metes and bounds states, the assumption is magnetic).
- Directions and distances are to be run in the original order of establishment.
- Distance over direction except in metes and bounds states where direction over distance.
- Distances are assumed ground distances.

Areas

Area calls are considered with the least weight and are utilized if no other element of construction is available.

RIPARIAN AND LITTORAL RIGHTS

Property rights associated with water are known riparian and littoral rights:

Littoral Rights: Pertain to shore frontages and lakes.

Riparian Rights: Pertain to rivers and streams.

Navigable: Able to be sailed by ships and/or boats.

Non-navigable: Not able to be sailed by ships and/or boats.

The classification of whether or not a body of water is navigable is dependent on each individual state's statutes. The definitions of navigable depend on a range of considerations (e.g., waterbody's depth, width, use, historic application, geographic location, and many other elements).

Sudden change vs. gradual change by action of water

Typically, land rights are altered with a gradual change.

Typically, land rights are not altered with a sudden change.

Terms

High-water line: The water level and average height of the water mark during high tide durations if tidal or typically during spring/summer months if non-tidal.

Mean high-water line: The height of water along a water body, which is the average of the higher water ranges.

Low-water line: The water level and average height of the water mark during low tide durations if tidal or typically during fall/winter months if non-tidal.

Mean low-water line: The height of water along a water body, which is the average of the lower water ranges.

Shore: Tidelands affected between mean high-water line and mean-low water line.

Bank: The line along rivers and streams at which point the vegetation line does not encroach.

Beach: The area between the mean-low water line and the line of vegetation/material change.

Thalweg: The deepest part of the deepest channel of a river or stream.

Thread: Sometimes used interchangeably with the term thalweg, but in certain states it is defined as the actual middle of the river or stream being a line equidistant from each bank.

Avulsion: The sudden loss or gain of land by the action of water. Typically, land rights are not altered with a sudden change.

Erosion: The gradual loss of land by the action of water. Typically, land rights are altered with a gradual change.

Accretion: The gradual addition of land by the action of water. Typically, land rights are altered with a gradual change.

Alluvion/Alluvium: The actual material/soil or land that is built up and deposited during accretion.

Reliction: The gradual recession of water causing the exposure of previously submerged lands. Typically, land rights are altered with a gradual change.

Submergence: The gradual rise of water causing the covering of previously unsubmerged lands. Typically, land rights are altered with a gradual change.

Tidal Cycle: The duration of time necessary to completely analyze tidal levels is 18.6 years.

Note: As with navigability, water rights are dependent on a State's legislative actions and can vary greatly among states and regions. The terms herein are general in nature and attempt to outline the overall concepts of water rights and the associated boundaries.

ORDER OF OPERATIONS:

PARENTHESES, EXPONENTS, MULTIPLICATION, DIVISION, ADDITION, AND SUBTRACTION (PEMDAS)

Following the order of operations is essential to derive the correct answer. The order needs to be followed along with the movement through the equation from left to right.

1. parentheses
2. exponents/roots
3. multiplication
4. division — (Note: Multiplication and division are equal in priority)
5. addition
6. subtraction — (Note: Addition and subtraction are equal in priority)

Algebra Fundamentals

- Solve for the unknown/variable; Variable = "___" = "?" = "x".
- "=" is a mirror; what is done to one side of equation must be done to the opposite side.
- The goal is to isolate the variable to one side.
- Isolation requires reduction; reduction = negation (i.e., subtraction of positive numbers, addition of negative numbers, divide for multiplication, and multiply for division, opposite operation for negation).
- Remember 1) PEMDAS and 2) moving left to right.

Examples of PEMDAS

Example 1

$2 + 6 \times 3 =$	VS.	$2 + 6 \times 3 =$
$2 + 18 =$		$8 \times 3 =$
$\mathbf{20}\ (\boldsymbol{correct})$		$\mathbf{24}\ (\boldsymbol{wrong})$
$30 \div 5 \times 2 =$	VS.	$30 \div 5 \times 2 =$
$6 \times 2 =$		$30 \div 10 =$
$\mathbf{12}\ (\boldsymbol{correct})$		$\mathbf{3}\ (\boldsymbol{wrong})$

Example 2

$\sqrt{25}(5+2)^2 - 18 \div 3(3-1) + 2^3 =$	perform operation in parenthesis first
$\sqrt{25}(7)^2 - 18 \div 3(2) + 2^3 =$	perform all exponent and root operations
$5(49) - 18 \div 3(2) + 8 =$	perform all multiplication/division (moving L to R)
$245 - 18 \div 3(2) + 8 =$	
$245 - 6(2) + 8 =$	
$245 - 12 + 8 =$	perform all addition/subtraction (moving L to R)
$168 + 8 =$	
$\mathbf{241}$	

Example 3

$[(5-6)^2+3\times 4-6]\times 2=$ parenthesis

$[(-1)^2+3\times 4-6]\times 2=$ exponents

$[1+3\times 4-6]\times 2=$ multiplication inside parenthesis

$[1+12-6]\times 2=$ addition and subtraction inside parenthesis (moving L to R)

$[7]\times 2=$

$7\times 2=$

$\mathbf{14}=$

Example 4

$1-15\times 2+25-3+15-3=$

$1-30+25-3+15-3=$

$-29+25-3+15-3=$

$-4-3+15-3=$

$-7+15-3=$

$\mathbf{5}$

Example 5

$y-4=\frac{y}{4}+\frac{1}{8}$

$\frac{8}{1}(y-4)=\left(\frac{y}{4}+\frac{1}{8}\right)\frac{8}{1}$ multiple each side by the inverse of 1/8 (8) to negate 1/8

$8y-32=\frac{8y}{4}+\frac{8}{8}$ distribute 8/1 into each parenthesis on each side

$8y-32=2y+1$

$\underline{\quad +32 \qquad +32}$ add 32 to each side

$8y=2y+33$

$\underline{-2y\quad -2y}$ subtract 2*y* from each side

$6y=33$

$\frac{6y}{6}=\frac{33}{6}$

$y=\mathbf{5.5}$

Exponent rules

Product:	$a^x \times a^y = a^{x+y}$	When multiplication of exponents occurs ADD the exponents
Quotient:	$a^x \div a^y = a^{x-y}$	When division of exponents occurs SUBTRACT the exponents
Power:	$(a^x)^y = a^{xy}$	When powers of exponents occurs MULTIPLY the exponents
Zero:	$a^0 = 1$	Anything with an exponent of zero equals ONE
Negative:	$a^{-x} = 1/a^x$	Negative exponents call for an INVERSE

Examples of exponent rules

Example 1

$(3x)^2 = 3^2 \times x^{1\times 2}$

$= 9 \times x^2$

$= \mathbf{9x^2}$

Example 2

$x^3 - 2x^2(x+1) =$ distribute $2x^2$ using exponent rules

$x^3 - 2x^3 - 2x^2 =$ combine like variables (combine the x^3 $terms$)

$\mathbf{-x^3 - 2x^2} =$ equation is simplified

Simplification

Sometimes an answer cannot be derived; in this case, it is necessary to simplify the equation down to the minimal expression. Simplification is the combination of like variables and the reduction of like variables into their most reduced/simplified form.

Simplification examples include: (15/3 = 5), (18/36 = 1/2), or ($9n+3n+n^2 = 12n+n^2$).

Examples of simplification

Example 1

$5x^2y^3 - (6xy - 4x^2y^3 + 3) + 6 =$

$5x^2y^3 - 6xy + 4x^2y^3 - 3 + 6 =$ distribute the "-"

$5x^2y^3 - 6xy + 4x^2y^3 - 3 + 6 =$ combine the x^2y^3 $terms$

$\mathbf{9x^2y^3 - 6xy + 3}$

Example 2

$5xy - y(6x - 2x^2) =$

$4xy - 6xy + 2x^2y =$ distribute $-y$ and combine xy $term$

$\mathbf{-2xy + 2x^2y} =$

Example 3

$(6x^2y)^3 =$

$\mathbf{216x^6y^3} =$

QUADRATIC EQUATIONS

QUAD means squared: a squared variable "x^2" must be present. Quadratic equations are used to find the curves on a Cartesian grid and can be used practically to find the curve of an object in flight (e.g., as the speed of a ball in flight approaches zero the peak of flight is reached and then curves back down to the ground as its speed increases).

Quadratic equation:

$$ax^2 + bx + c = 0$$

or

Roots form of quadratic equation:

$$x = \frac{-b \pm \sqrt{b^2 - 4ac}}{2a}$$

Discriminate portion

$$b^2 - 4ac$$

If the discriminate is (+) positive, then 2 real answers are possible.
If the discriminate is (0) zero, then 1 real answer is possible.
If the discriminate is (-) negative, then complex solutions occur where: $(i = \sqrt{-1})$ or $(i^2 = -1)$.

"a" designation

If (+) positive, then "u" shaped parabola.
If (-) negative, then "n" shaped parabola.

The larger the value the more squeezed, and the smaller the value the wider the shape.

"b" designation

If (+) positive, then parabola's vertex is on the negative side of x-axis.
If (-) negative, then parabola's vertex is on the positive side of x-axis.
If (0) zero, then parabola's vertex is centered on x-axis.

"c" designation

Y-intercept, where the parabola crosses the y-axis.

Vertex coordinates

In order to find the vertex coordinate values, solve the following equations.

$$h = \frac{-b}{2a}$$

Where: $h = x$ value

$$k = a(h)^2 - b(h) + c$$

Where: $k = y$ value

Steps of solving quadratic equation

Step 1: Read the quadratic equation to find out all parabola characteristics. The parabola direction a, which side of the x-axis the vertex is on b, and were does the parabola cross the y-axis; c.

Step 2: Solve the discriminate to figure out if 1 (if equals 0), 2 (if equals +), or complex (if equals -) solutions are possible.

Step 3: Solve for x values with the roots form of the quadratic equation.

Step 4: Solve for vertex x,y coordinates. This is performed by the h and k value equations.

Step 5: Check by inserting in the found values from the roots form equation into the original equation. If values equal zero, then it is correct.

Example 1

$x^2 + 3\text{x} - 4 = 0$

Step 1: Characteristics

a is positive, so  shaped

b is positive, so parabola vertex will be on negative side of x-axis

c is -4, so parabola crosses y-axis a (0,-4)

Step 2: Discriminate

$b^2 - 4ac =$

$3^2 - 4(1)(-4) =$

$9 + 16 =$

$\mathbf{25}$; the discriminate is positive, so 2 real number solutions are possible

Step 3: Solutions derived from roots form of quadratic equation.

$$x = \frac{-b \pm \sqrt{b^2 - 4ac}}{2a}$$

$$x = \frac{-3 \pm \sqrt{3^2 - 4(1)(-4)}}{2(1)}$$

$$x = \frac{-3 \pm \sqrt{25}}{2}$$

$$x = \frac{-3 \pm 5}{2}$$

$$x = \frac{-3 + 5}{2} = \frac{2}{2} = \mathbf{1}$$

and

$$x = \frac{-3 - 5}{2} = \frac{-8}{2} = \mathbf{-4}$$

Step 4: Solve for vertex coordinates.

$$h = \frac{-b}{2a}$$

$$h = \frac{-3}{2(1)}$$

$$h = \mathbf{-1.5} = \boldsymbol{x}$$

$$k = a(h)^2 - b(h) + c$$

$$k = 1(1.5)^2 - 3(1.5) + (-4)$$

$$k = 2.25 - 4.5 - 4$$

$$k = \mathbf{-6.25} = \boldsymbol{y}$$

Step 5: Check found x values from roots form in original quadratic equation.

One x solution was **1** so

$x^2 + 3\text{x} - 4 = 0$

$\mathbf{1}^2 + 3(\mathbf{1}) - 4 = 0$

$1 + 3 - 4 = 0$

$0 = 0$ ***check***

One x solution was **-4** so

$x^2 + 3\text{x} - 4 = 0$

$\mathbf{-4}^2 + 3(\mathbf{-4}) - 4 = 0$

$16 - 12 - 4 = 0$

$0 = 0$ ***check***

Example 2

$5x^2 + 2\text{x} + 1 = 0$

Step 1: Characteristics

a is positive, so  shaped

b is positive, so parabola vertex will be on negative side of x-axis

c is 1, so parabola crosses y-axis a (0,1)

Step 2: Discriminate

$b^2 - 4ac$

$2^2 - 4(5)(1)$

$4 - 20$

$\mathbf{-16}$; the discriminate is negative, so no real number solutions are possible (meaning the parabola will not cross the x-axis)

Step 3: Solutions derived from roots form of quadratic equation

$$x = \frac{-b \pm \sqrt{b^2 - 4ac}}{2a}$$

$$x = \frac{-2 \pm \sqrt{2^2 - 4(5)(1)}}{2(5)}$$

$$x = \frac{-2 \pm \sqrt{-16}}{10}$$

$$x = \frac{-2 \pm 4\text{i}}{10}$$

$$x = \frac{-2 + 4\text{i}}{10} = \mathbf{-0.2 + 0.4}\boldsymbol{i}$$

and

$$x = \frac{-2 - 4\text{i}}{10} = \mathbf{-0.2 - 0.4}\boldsymbol{i}$$

The complex solutions are "not real", the solutions are called imaginary *i*. Imaginary numbers cannot be graphed, but the mirror image of the solutions can be. In other words, if the  shape is inverted to the  shape, then the solutions would intersect the x-axis. To "flip the shape" simply remove the *i* for the solution and solve. In this case, it would be ($\mathbf{-0.2 + 0.4}\boldsymbol{i}$; remove *i* and get $\mathbf{-0.2 + 0.4 = 0.2}$) and ($\mathbf{-0.2 - 0.4}\boldsymbol{i}$; remove *i* and get $\mathbf{-0.2 - 0.4 = -0.6}$). Meaning the parabola of the inverted shape would intersect the x-axis at -0.6 and 0.2.

Step 4: Solve for vertex coordinates.

$$h = \frac{-b}{2a}$$

$$h = \frac{-2}{2(5)}$$

$$h = \mathbf{-0.2} = \boldsymbol{x}$$

$$k = a(h)^2 - b(h) + c$$

$$k = 5(-0.2)^2 - 2(-0.2) + (1)$$

$$k = 0.2 - 0.4 + 1$$

$$k = \mathbf{0.8} = \boldsymbol{y}$$

Step 5: Check found x values from roots form in original quadratic equation.

One x solution was **-0.2 + 0.4i** so;

$5(-0.2+0.4i)^2+2(-0.2+0.4i)+1=0$

$5(-0.2+0.4i)(-0.2+0.4i)+2(-0.2+0.4i)+1=0$

$5(0.04-0.08i-0.08i+0.16i^2)+2(-0.2+0.4i)+1=0$

$5(0.04-0.16i+0.16(-1))+2(-0.2+0.4i)+1=0$ **remember;**$(\boldsymbol{i}=\sqrt{-1})$ **or** $(\boldsymbol{i^2}=-\mathbf{1})$

$5(-0.12-0.16i)+2(-0.2+0.4i)+1=0$

$-0.6-0.8i+(-0.4)+0.8i+1=0$

$-0.6-0.4+1=0$

$-1+1=0$

$0=0\ \boldsymbol{check}$

The other x solution was **-0.2 - 0.4i**; this solution would not work to allow for the
 shape to work. Therefore, a check is not necessary.

TRIANGLES:
RIGHT AND OBLIQUE TRIANGLES
PYTHAGOREAN THEOREM, LAW OF SINES, AND THE LAW OF COSINES

Trigonometry is the mathematical study of the relationship of side lengths and angle measurements associated with triangular figures. Trigonometric study can be further divided into right-angled triangular figures and oblique triangular figures.

Information about triangles

- The summation of the interior angles of any triangle will equal 180°.
- Obtuse triangles have one angle over 90°.
- Acute triangles have no angle over 90°.
- A right triangle is a triangle having one of the angles equaling 90°.
- An oblique triangle does not have a right angle.
- Equilateral triangles have three equal length sides.
- Scalene triangles have no equal length sides.
- Isosceles triangles have two equal length sides.
- Right triangle solutions are found using the Pythagorean Theorem, Sine, Cosine, and Tangent operations.
- The LAW OF SINES and the LAW OF COSINES apply only to NON-right triangles.

Right triangles

Solving for unknown side lengths and angles of right triangles.

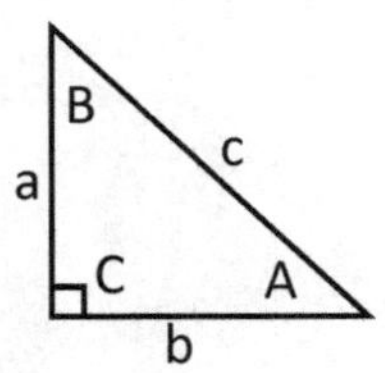

Where: "c" is the hypotenuse and "C" is 90°

Pythagorean Theorem: $a^2 + b^2 = c^2$ or $c = \sqrt{a^{2+}b^2}$

The concept behind the Pythagorean Theorem is the summation of the areas created by squares subtended from the two non-hypotenuse legs of a right triangle equal the area of a square subtended from the hypotenuse leg (see image below).

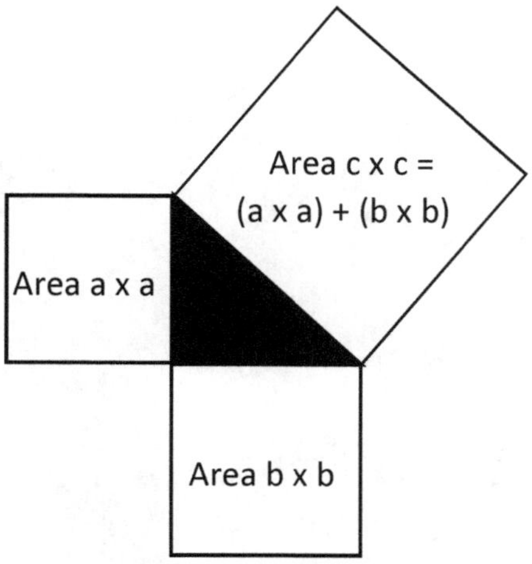

<u>Solutions for unknown elements of a right triangle are found with Sine, Cosine, and Tangent</u>

Name	abbreviation	operation	rhyme
Sine	SIN	$\frac{opposite}{hypotenuse}$	"SOH"
Cosine	COS	$\frac{adjacent}{hypotenuse}$	"CAH"
Tangent	TAN	$\frac{opposite}{adjacent}$	"TOA

<u>Reciprocals of Sine, Cosine, and Tangent are known as Cosecant, Secant, and Cotangent respectively</u>

Name	abbreviation	operation	reciprocals
Cosecant	CSC	$\frac{hypotenuse}{opposite}$	1/SIN or (SIN^{-1})
Secant	SEC	$\frac{hypotenuse}{adjacent}$	1/COS or (COS^{-1})
Cotangent	COT	$\frac{adjacent}{opposite}$	1/TAN or (TAN^{-1})

Oblique triangles

Solutions for oblique triangles are found through the use of LAW OF SINES and the LAW OF COSINES. The laws solve for unknown side lengths and angles of oblique triangles.

Note the "pairs"
angle "A" and its opposing side "a"
angle "B" and its opposing side "b"
angle "C" and its opposing side "c"

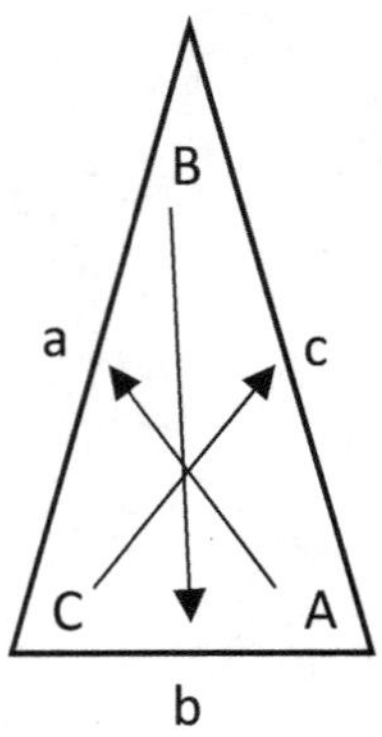

Law of Sines

$$\frac{a}{sinA} = \frac{b}{sinB} = \frac{c}{sinC}$$

The concept of the law of Sines is the ratio of any side's length divided by the Sine of that side's opposing angle will equal the ratio of any other side's length divided by the Sine of its opposing angle.

Law of Cosines

$$c^2 = a^2 + b^2 - 2ab\ (cosC)$$ or $$cosA = \frac{b^2+c^2-a^2}{2bc}$$

Variations of Law of Cosines for all three angles:

$$cosA = \frac{b^2 + c^2 - a^2}{2bc}$$

$$cosB = \frac{a^2 + c^2 - b^2}{2ac}$$

$$cosC = \frac{a^2 + b^2 - c^2}{2ab}$$

Variations of Law of Cosines for all three sides:

$$a^2 = b^2 + c^2 - 2bc\ (cosA)$$

$$b^2 = a^2 + c^2 - 2ac\ (cosB)$$

$$c^2 = a^2 + b^2 - 2ab\ (cosC)$$

Types of triangles and the associated solving methods (Note A=angle and S=side)

1) AAA		Angle Angle Angle

all three angles are known; and no sides are known

- Impossible to solve for sides, but summation of angles will equal 180°

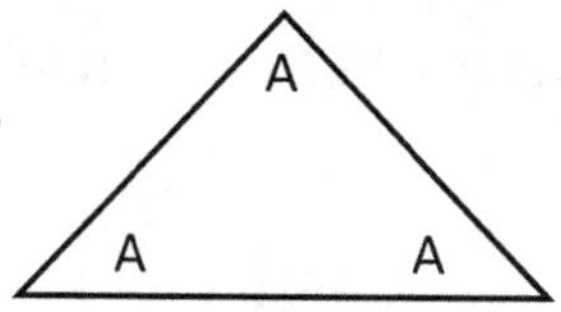

2) AAS		Angle Angle Side

two known angles and one known side (not between the known angles)

- First: summation of angles for missing angle
- Second: Law of Sines for missing sides

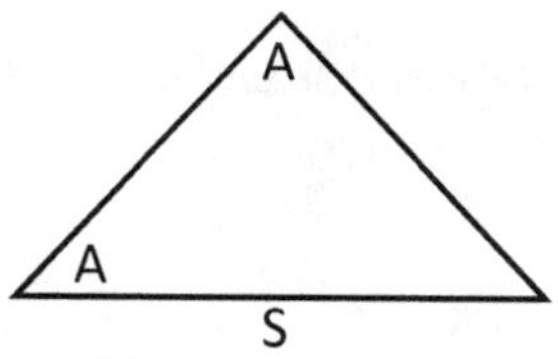

3) ASA		Angle Side Angle

two known angles and one known side (between two known angles)

- First: summation of angles for missing angle
- Second: Law of Sines for missing sides

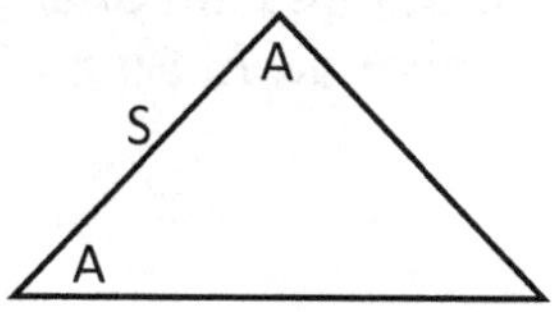

4) SAS		Side Angle Side

two sides and their inclusion angle are known

- First: Law of Cosines for the missing side
- Second: Law of Sines for first missing angle
- Third: summation of angles for last missing angle

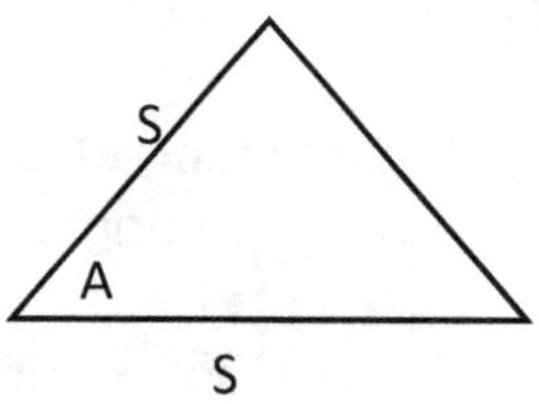

5) SSA		Side Side Angle

two sides and a non-inclusion angle are known

- First: Law of Sines for first missing angle
- Second: summation of angles for last missing angle
- Third: Law of Sines for last missing side

Note: See section below for SSA having two solutions

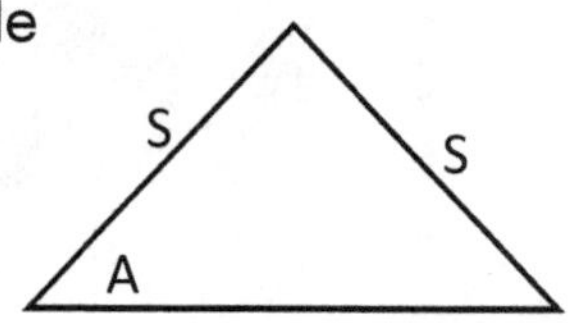

6) SSS		Side Side Side

all three sides and no angles are known

- First: Law of Cosines for first missing angle
- Second: Law of Sines or Cosines for second missing angle
- Third: summation of angle for final missing angle

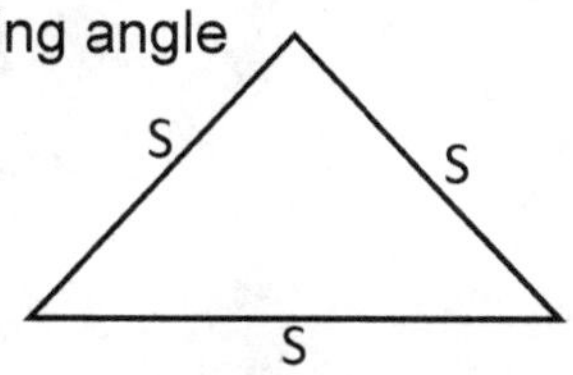

Example 1

Solve for all unknowns (a, A, and c) with knowns (b=7, B=35°, and C=105°).

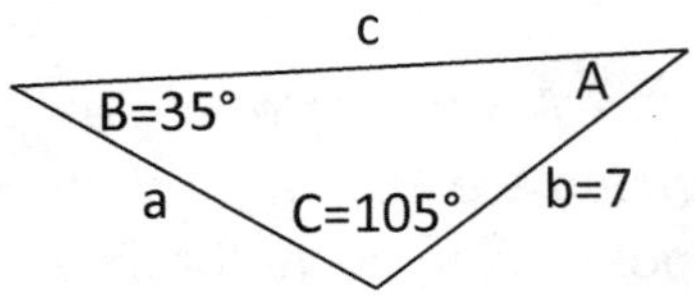

Note: Remember to utilize calculator memory operations to allow for calculating the accurate answer. When using Sine, Cosine, and Tangent long decimal numbers are produced and may alter the answer if significant numbers are reduced.

Note: When possible utilize provided information as oppose to solved information. This will reduce error.

Step 1: Identify the type of triangle.
The triangle type is AAS and method of solving is as follows:

First: Summation of angles for missing angle.
Second: Law of Sines for missing sides.

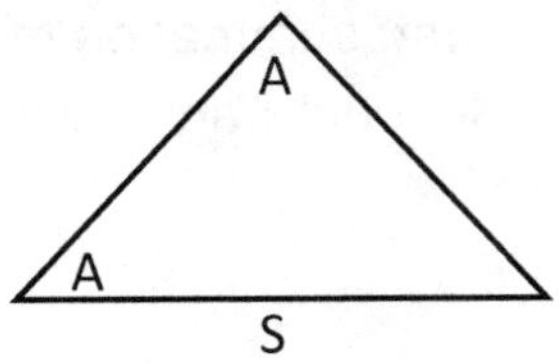

Step 2: Solve the unknown angle with the summation of angles.

The summation of angles in any triangle must be 180°;
We know two angles B=35° and C=105°, so the third is solved as follows:

180° = A + 35° + 105°
180° = A + 140°
40° = A

Step 3: Solve for the first unknown side with the Law of Sines.

Choose a "pair" (an angle and its opposing side) of knowns. Then, choose a "pair" (an angle and its opposing side) of which only one of the "pair" is known.

Law of Sines: $\frac{a}{sinA} = \frac{b}{sinB} = \frac{c}{sinC}$

$$\frac{a}{sinA} = \frac{b}{sinB}$$

$$\frac{a}{sin40°} = \frac{7}{sin35°}$$

$$\frac{a}{0.64278761} = \frac{7}{0.573576436}$$

$$\frac{a}{0.64278761} = 12.20412757$$

$$a = 12.20412757 \times 0.64278761$$

$$\boldsymbol{a = 7.844661992}$$

Step 4: Solve for the last unknown side again with Law of Sines.

Choose a "pair" (an angle and its opposing side) of knowns. Then, choose a "pair" (an angle and its opposing side) of which only one of the "pair" is known.

Law of Sines: $\frac{a}{sinA} = \frac{b}{sinB} = \frac{c}{sinC}$

$$\frac{a}{sinA} = \frac{c}{sinC}$$

$$\frac{7.844661992}{0.64278761} = \frac{c}{sin105°}$$

$$12.20412757 = \frac{c}{0.965925826}$$

$$12.20412757 \times 0.965925826 = c$$

$$\boldsymbol{11.78828201 = c}$$

Example 2

Solve for all unknowns (A, B, and C) with knowns are (a=8, b=6, and c=7).

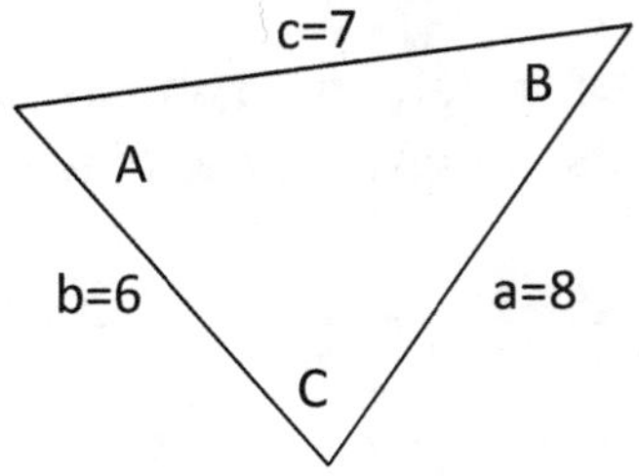

Step 1: Identify the type of triangle.
The triangle is SSS and method of solving is as follows:

First: Law of Cosines for first missing angle.
Second: Law of Sines or Cosines for second missing angle.
Third: Summation of angle for final missing angle.

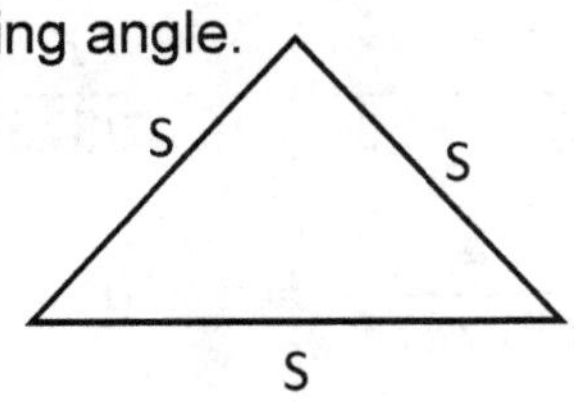

Step 2: Solve for the first unknown angle with the Law of Cosines.

Law of Cosines: $c^2 = a^2 + b^2 - 2ab\,(cosC)$ or $cosA = \frac{b^2+c^2-a^2}{2bc}$

$$cosA = \frac{b^2 + c^2 - a^2}{2bc}$$

$$cosA = \frac{6^2 + 7^2 - 8^2}{2(6)(7)}$$

$$cosA = \frac{36 + 49 - 64}{84}$$

$$cosA = \frac{21}{84}$$

$$cosA = 0.25$$

$$(cos^{-1})cosA = 0.25(cos^{-1})$$

$$\mathbf{A = 75.52248781}$$

Step 3: Solve the next unknown angle again with Law of Cosines.

Law of Cosines: $c^2 = a^2 + b^2 - 2ab\,(cosC)$ or $cosA = \frac{b^2+c^2-a^2}{2bc}$

$$cosB = \frac{a^2 + c^2 - b^2}{2ac}$$

$$cosB = \frac{8^2 + 7^2 - 6^2}{2(8)(7)}$$

$$cosB = \frac{64 + 49 - 36}{112}$$

$$cosB = \frac{77}{112}$$

$$cosB = 0.6875$$

$$(cos^{-1})cosB = 0.6875(cos^{-1})$$

$$\mathbf{B = 46.56746344}$$

Step 4: Solve the last unknown angle with the summation of angles.

The summation of angles in any triangle must be 180°;
We know two angles A=75.52° and B=46.57°, so the third is solved as follows:

180° = 75.52° + 46.57° + C
180° = 122.09° + C
57.91° = C

SSA: (Side Side Angle) triangles can have two separate and equally correct solutions. This does not occur with any other triangle type.

The triangle type of SSA presents a caveat of having two solutions possible. See the image below for illustration.

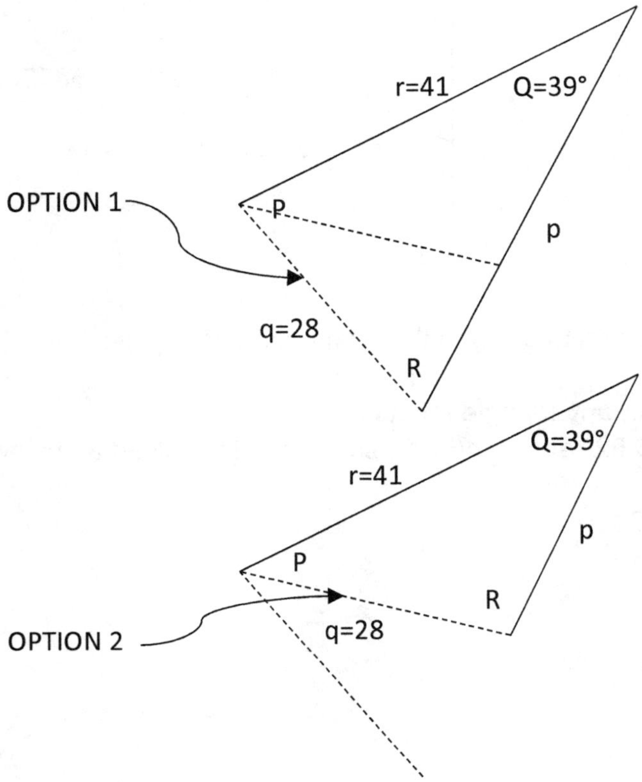

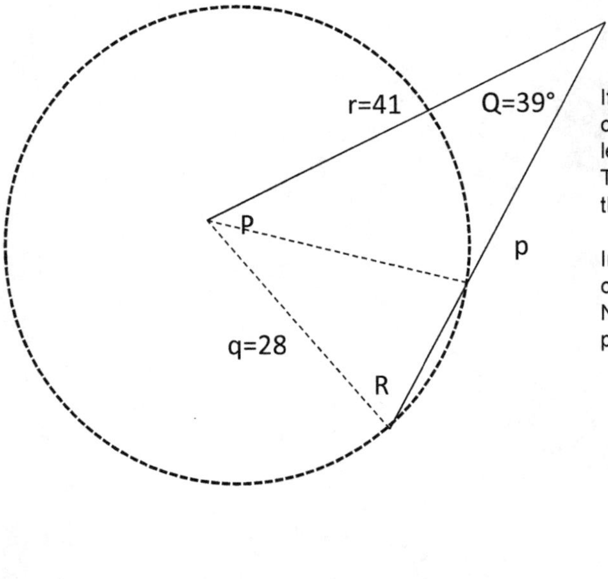

If the inclusion angle is held as the center of a circle, then it is illustrated that the "q" length of 28 is true in both options 1 and 2. The "inclusion" angle is the angle between the two known sides.

In this situation "R" can equal 67.14° or 112.86°; if the Law of Sines is applied. Note: the relationship of these two possible angles; 67.14°=180°-112.86°.

DECIMAL DEGREES AND DEGREES MINUTES SECONDS: CONVERSIONS

The manual process of converting:

degrees minutes second (DMS) to decimal degrees (DD)
decimal degrees (DD) to degrees minutes seconds (DMS)

DMS to DD example

Equation: $(\frac{seconds}{3600}+\frac{minutes}{60}+\frac{degrees}{1})$ = decimal degrees

<u>Example</u> Given 12° 43' 56.6"

Step 1: Convert seconds to decimal with the formula $(\frac{seconds}{3600})$. 56.6" divided by 3600 (a second is 1/3600th of a degree).

$$\frac{56.6''}{3600} = \mathbf{0.015722222}$$

Step 2: Convert minutes to decimal with the formula $(\frac{minutes}{60})$. 43' divided by 60 (a minute is 1/60th of a degree).

$$\frac{43'}{60} = \mathbf{0.716666}$$

Step 3: Convert degrees to decimal with the formula $(\frac{degrees}{1})$. 12° divided by 1:

$$\frac{12^\circ}{1} = \mathbf{12}$$

Finally: Add up all the results from the three steps.

$$\mathbf{0.015722222 + 0.716666 + 12 = 12.73238888^\circ}$$

DD to DMS example

In decimal degree conversion to degree minute second, the integer values represent the degree, minute, and second values and the decimal is further divided by 60.

<u>Example</u>

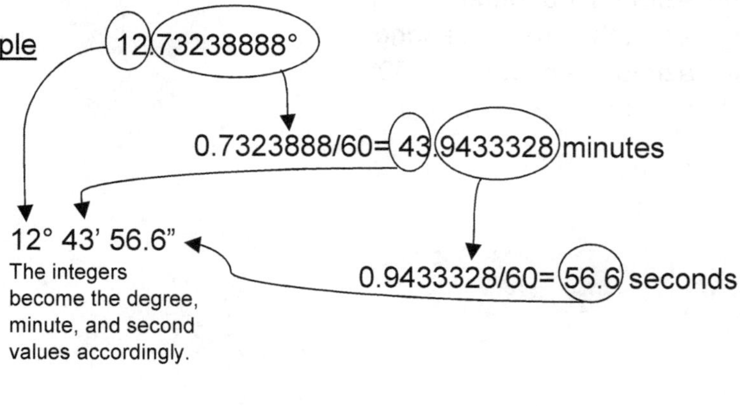

AZIMUTHS AND BEARINGS: CONVERSIONS

The ability to convert azimuths to bearings and bearings to azimuths is essential for survey calculations. The diagram below depicts the relation between the two types of angle measurements.

Azimuth and bearing diagram

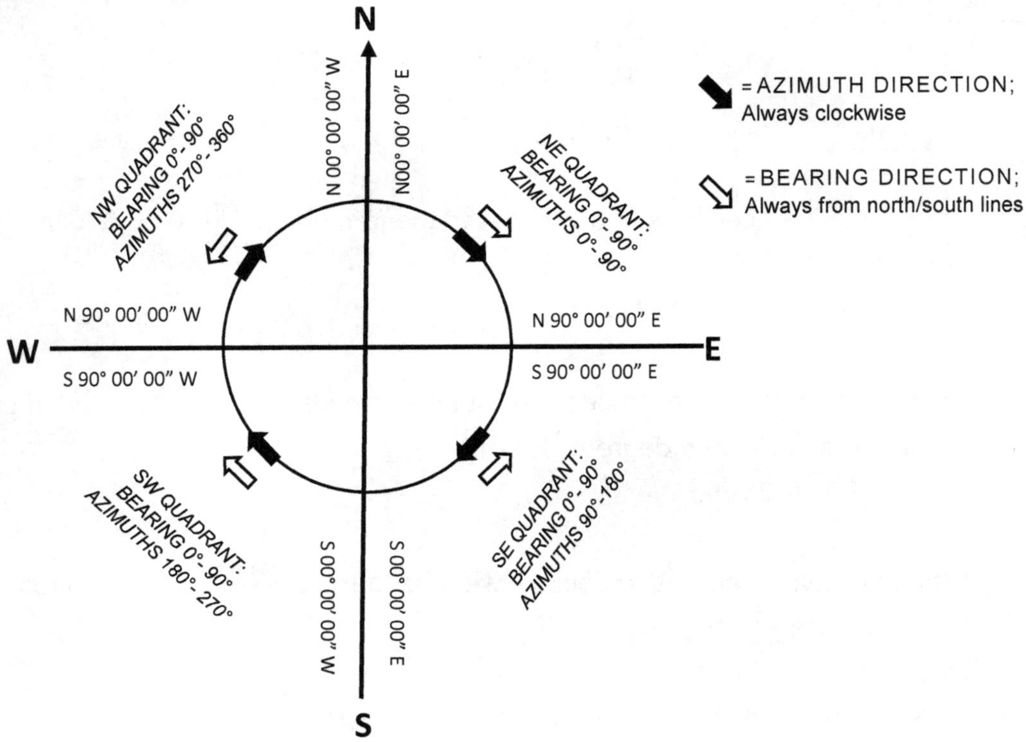

Conversion chart

For NE quadrant:	azimuth = bearing
For SE quadrant:	azimuth = 180° - bearing
For SW quadrant:	azimuth = bearing + 180°
For NW quadrant:	azimuth = 360° - bearing

MAGNETIC DECLINATION:

Magnetic declination is the angle measurement between True North and Magnetic North measured on a horizontal plane.

True North: The direction along meridians (lines of longitude) towards Geographic North.

Magnetic North: The direction of Earth's magnetic fields, shown by a needle of a compass. Magnetic North is always changing and is variable across Earth's surface.

True North and Magnetic North are different and the angle measurement between them is the declination.

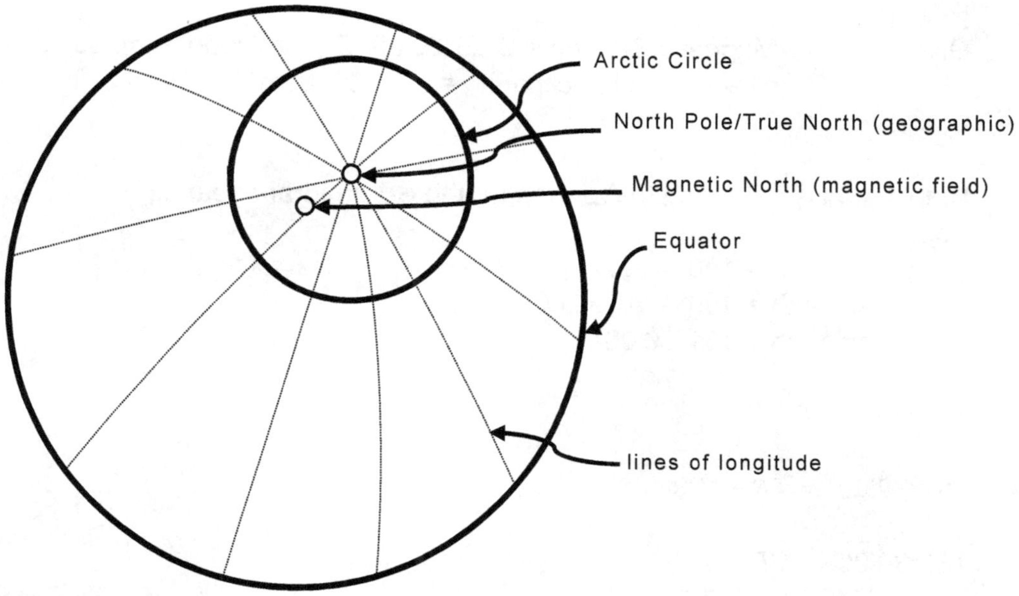

Declination equation

$$MB = TB + D$$
$$Magnetic\ Bearing = True\ Bearing + Declination$$

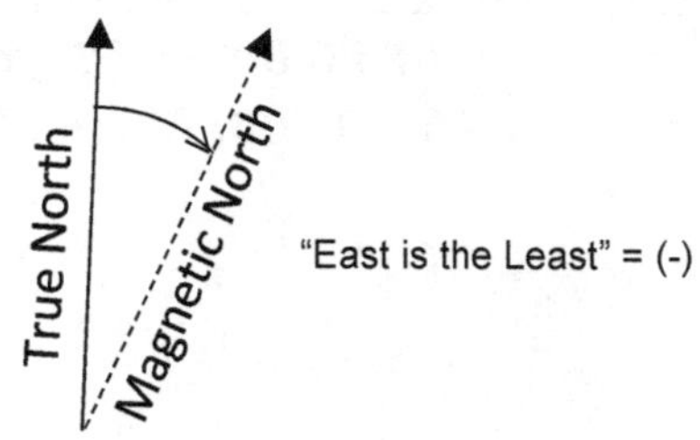

Where Declination D is:

(-) if Easterly shift from True North to Magnetic North; *"east is the least"*

(+) if Westerly shift from True North to Magnetic North; *"west is the best"*

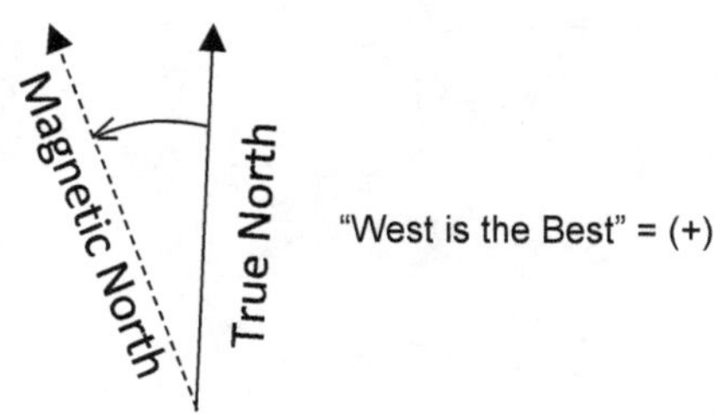

Example 1

Given: Magnetic Bearing = Az. 48°24'00" (in azimuth form so, no conversion needed) and declination = 5°38'00" East

Unknown: True Bearing

$MB = TB + D$

$48°24'00" = TB + (-5°38'00")$

$48°24'00" = TB - 5°38'00"$
$\underline{+5°38'00" \qquad + 5°38'00"}$
$\mathbf{54°02'00" = TB\ (true\ bearing)}$

Example 2

Given: Magnetic Bearing = S 28°30'00" E (in bearing form, so conversion is needed) and declination = 7°30'00" West

Unknown: True Bearing

Step 1: convert Bearing to Azimuth for the SE quadrant bearing.
For SE quadrant;

azimuth = 180° - bearing;
azimuth = 180° - 28°30'00"
azimuth = 151°30'00"

$MB = TB + D$

$151°30'00" = TB + 7°30'00"$

$151°30'00" = TB + 7°30'00"$
$\underline{-7°30'00" \qquad - 7°30'00"}$
$azimuth\ \mathbf{144°00'00" = TB\ (true\ bearing)}$
$bearing\ \boldsymbol{S}\ \mathbf{35°00'00"\ E = TB\ (true\ bearing)}$

AREA, VOLUME, AND CIRCUMFERENCE EQUATIONS

Curve areas

$$sector = \frac{r \times L}{2} = \frac{\pi r^2 I}{360°}$$

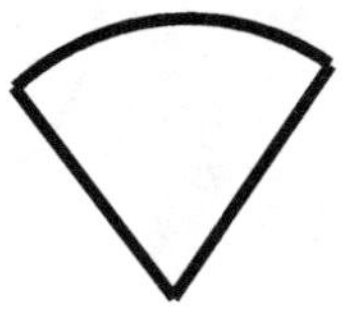

$$segment = \frac{\pi r^2 I}{360°} - \frac{r^2 sin I}{2}$$

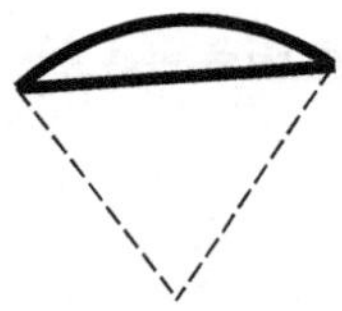

$$fillet = r\left(T - \frac{L}{2}\right)$$

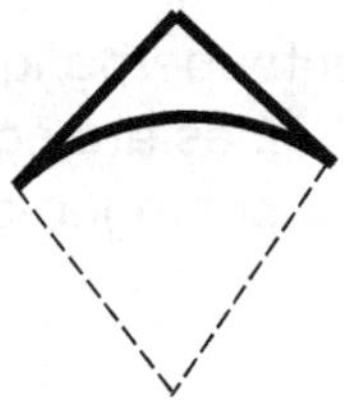

Note: Refer to horizontal curve section for variable designations and definitions.

Triangle areas

$$right\ triangle = \frac{1}{2} base \times height$$

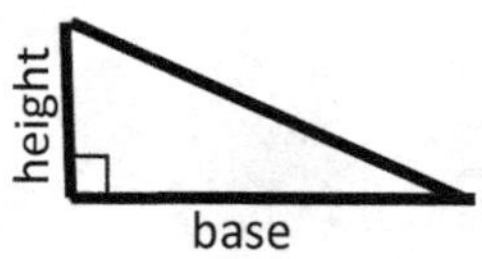

$$oblique\ triangle = \frac{a\ b\ sinC}{2} = \frac{a^2\ sinB\ sinC}{2\ sinA}$$

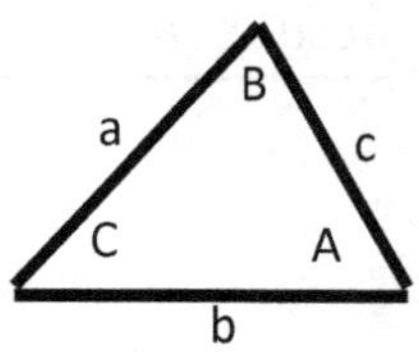

Circle area and circumference

$$circle\ area = \pi r^2$$

$$circle\ circumference = 2\pi r$$

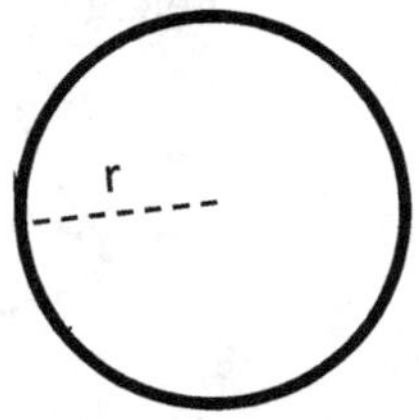

Sphere volume

$$sphere\ area = 4\pi r^2$$

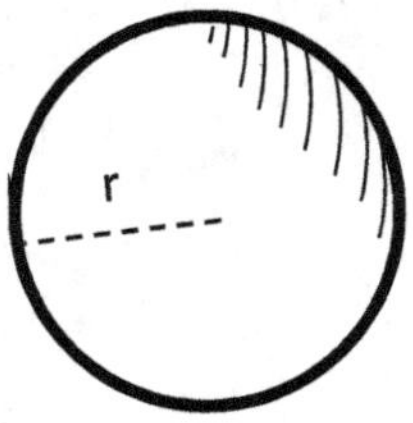

Polygon areas (assumes all sides are equal)

$$area\ of\ polygon = \frac{pLn}{2}$$

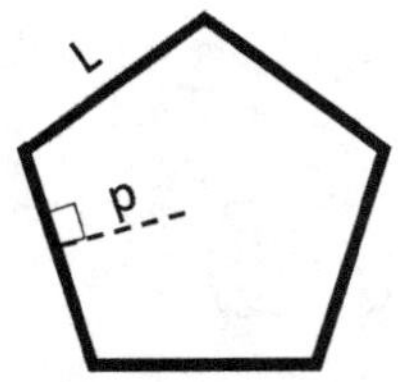

Where: *p*=length of line from center perpendicular to any side
L=length of any side (all sides are equal)
n= number of sides (e.g., pentagon-5, hexagon-6, etc.)

$$area\ of\ polygon\ with\ known\ diagonals = \frac{(sin\alpha)(d_1)(d_2)}{2}$$

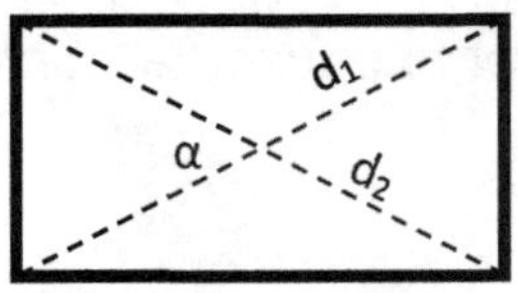

Where: α= smaller angle produced by the intersecting diagonals.

Pyramid or cone volumes

$$volume\ of\ pyramid/cone = \frac{height\ (area\ of\ base)}{3}$$

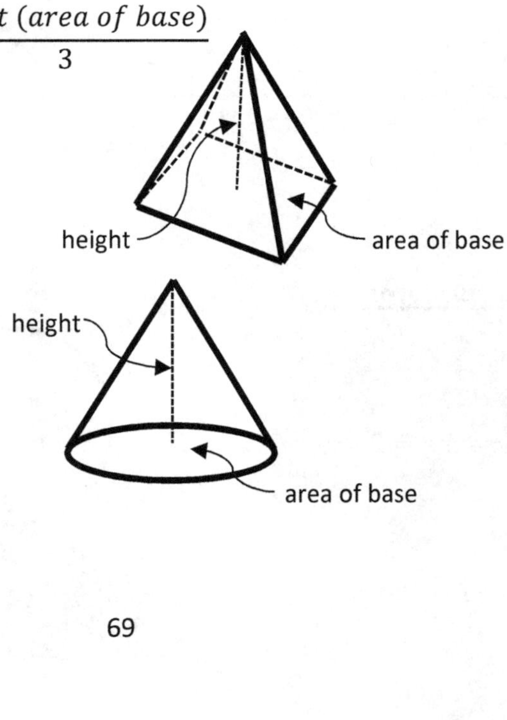

Trapezoid area

$$trapezoid\ area = \left(\frac{base_1 + base_2}{2}\right) altitude$$

Where: $base_1$ and $base_2$ are parallel lines, and altitude is the perpendicular distance between $base_1$ and $base_2$

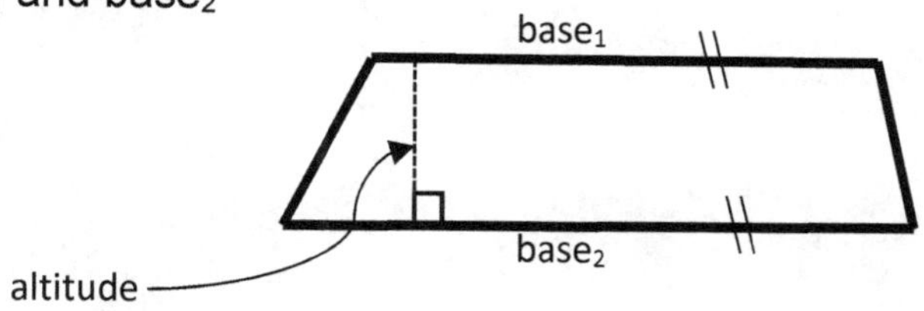

Prismoidal volume

$$prismoidal\ volume = \frac{length\ (A_1 + 4A_m + A_2)}{6}$$

Where: A_1 and A_2 are areas of the end (first and second) trapezoids. A_m is equal to $\frac{A_1+A_2}{2}$

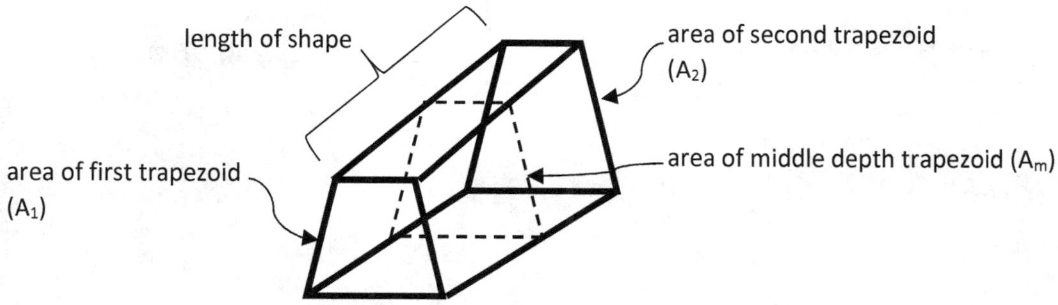

The prismoidal volume is the best method for volume calculations associated with earthwork.

Area of closed polygon from coordinates

$$closed\ polygon\ area = \left|\frac{1}{2}[(x_A y_B - x_B y_A) + (x_B y_C - x_C y_B) + (x_C y_D - x_D y_C) + (x_D y_A - x_A y_D)]\right|$$

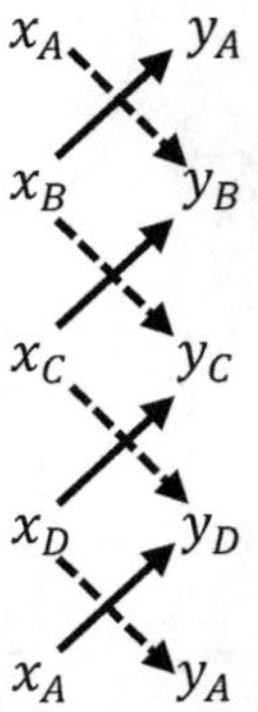

Note: Do not forget to incorporate the starting point at the end; this closes the polygon.

Example

	X	Y		
A	4	9		
			(4X7) – (9X9) =	-53
B	9	7		
			(9X2) – (11X7) =	-59
C	11	2		
			(11X2) – (2X2) =	18
D	2	2		
			(2X9) – (4X2) =	10
A	4	9	+	
				-84

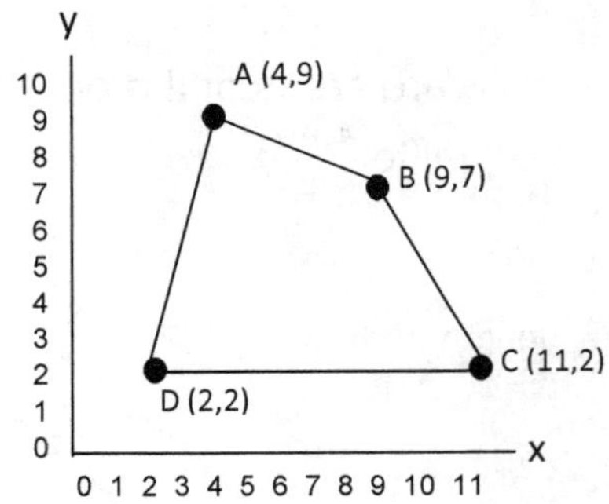

$$area\ of\ closed\ polygon = \left|\frac{1}{2}[-84]\right| = 42$$

Trapezoidal rule

The concept associated with the trapezoidal rule method of determining area is based on the summation of several trapezoids. The area being determined is the area below a graphed line within a Cartesian grid.

There are two methods of applying the trapezoidal rule. One is for **equally** spaced x values and one for **unequally** spaced x values.

equally spaced x values

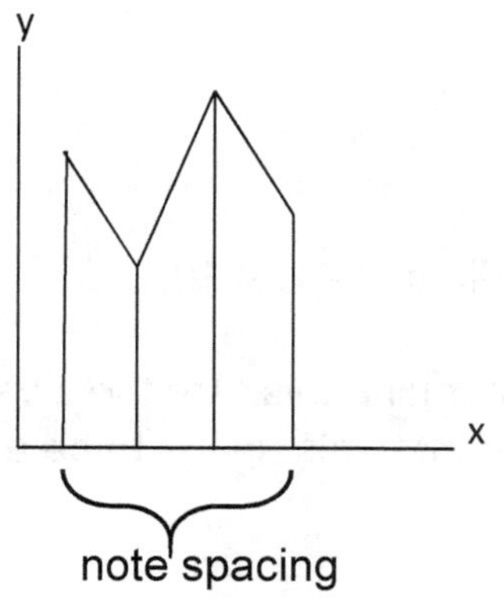

unequally spaced x values

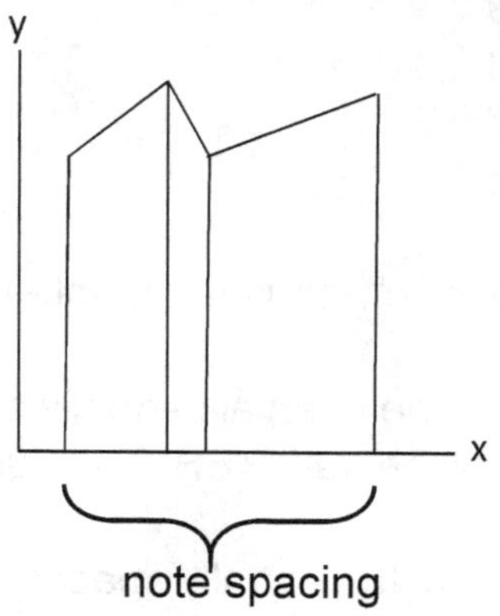

Equally spaced x value equation

$$area = h\left[\frac{1}{2}(y_0 + y_n) + y_1 + y_2 + y_3 + \cdots\right]$$

Where: $h = x_1 - x_0$

Unequally spaced x value equation

$$area = \frac{(y_0 + y_1)\Delta x}{2} + \frac{(y_1 + y_2)\Delta x}{2} + \frac{(y_2 + y_3)\Delta x}{2} + \cdots$$

If thought of in terms of "multiple trapezoids summation":

$$area = \frac{(base_1 + base_2)height}{2} + \frac{(base_2 + base_3)height}{2} + \frac{(base_3 + base_4)height}{2} + \cdots$$

Where: Base lengths are associated with the y values and the height lengths are associated with the x values.

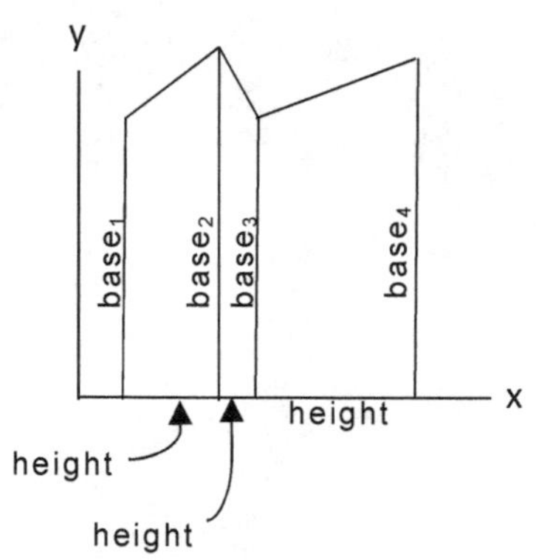

Example 1

Find the area of multi-trapezoid figure using the provided coordinate:

X	Y
0	4
1	3
2	4
3	7
4	12
5	20
6	28
7	39
8	53
9	67
10	84

Step 1: Find "n". "n" is the number of intervals (set of coordinates - 1) = 10.

Step 2: Are the intervals equally spaced? The x values determine this. Are the x values evenly incremental? YES. The equation for equally spaced intervals needs to be applied.

Step 3: Apply the equally spaced interval equation.

$$area = h\left[\frac{1}{2}(y_0 + y_n) + y_1 + y_2 + y_3 + \cdots\right]$$

Where: $h = x_1 - x_0$

Step 1:

$h = x_1 - x_0$

$h = 1 - 0 = 1$ (the x values for the first and second coordinate set)

then:

$$area = 1\left[\frac{1}{2}(4 + 84) + 3 + 4 + 7 + 12 + 20 + 28 + 39 + 53 + 67\right]$$

$$area = 1[44 + 233]$$

$$area = 277$$

Example 2

Find the area of multi-trapezoid figure using from the provided coordinate:

X	Y
1	10
3	12
4	16
7	20

Step 1: Find "n". "n" is the number of intervals (set of coordinates - 1) = 3.

Step 2: Are the intervals equally spaced? The x values determine this. Are the x values evenly incremental? NO. The equation for unequally spaced intervals needs to be applied.

Step 3: Apply the equally spaced interval equation.

$$area = \frac{(y_0 + y_1)\Delta x}{2} + \frac{(y_1 + y_2)\Delta x}{2} + \frac{(y_2 + y_3)\Delta x}{2} + \cdots$$

$$area = \frac{(10 + 12)(3 - 1)}{2} + \frac{(12 + 16)(4 - 3)}{2} + \frac{(16 + 20)(7 - 4)}{2}$$

$$area = 22 + 14 + 54$$

$$area = 90$$

Note: The resulting area found using trapezoidal rule method is to be considered approximate if a curved line is graphed on a Cartesian grid as opposed to straight lines.

Simpson's 1/3 Rule

When utilizing the trapezoidal rule, the area under the coordinates was found by connecting the points with straight lines. Simpson's 1/3 rule considers a curve between points as oppose to straight lines.

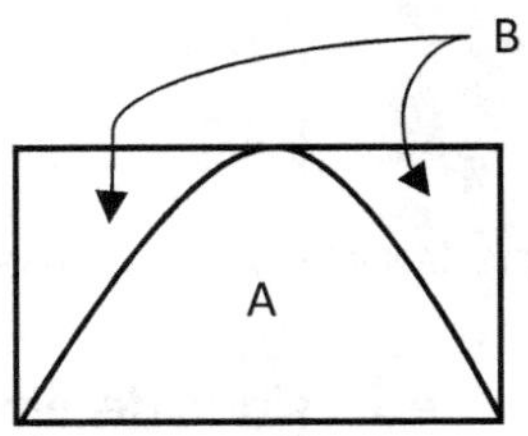

Concept: The area in "A" is 2/3 of the area of "B" in the exhibit.

Equation with the application of coordinate values.

$$area = \frac{\Delta x}{3}[y_0 + 4y_1 + 2y_2 + 4y_3 + 2y_4 + \cdots + y_n]$$

Where: $\Delta x = \frac{b-a}{n}$

b = largest x value
a = smallest x value
n = number of intervals (largest – smallest / number of intervals)

Example

Using the Simpson's 1/3 rule; what is the area below the curve on the Cartesian grid created by the provided coordinates?

X	Y
0	2
1	7
2	11
3	10
4	5

Step 1: Find the number of intervals (set of coordinates – 1) = 4.

n = 4

Step 2: Determine $\Delta x = \frac{b-a}{n}$

Where:

"b" = largest x value

"a" = smallest x value

"n" = number of intervals (largest – smallest / number of intervals)

$$\Delta x = \frac{b-a}{n}$$

$$\Delta x = \frac{4-0}{4}$$

$$\Delta x = 1$$

Step 3: Apply Simpson's 1/3 rule coordinate equation.

$$area = \frac{\Delta x}{3}[y_0 + 4y_1 + 2y_2 + 4y_3 + 2y_4 + \cdots + y_n]$$

$$area = \frac{1}{3}[2 + 4(7) + 2(11) + 4(10) + 5]$$

$$area = \frac{1}{3}[2 + 28 + 22 + 40 + 5]$$

$$area = \frac{1}{3}[97]$$

$$area = 32.333$$

HORIZONTAL CURVES

A major aspect of survey is the horizontal curve and solving for the different elements of a horizontal curve. This section outlines the procedures for solving various elements of horizontal curves.

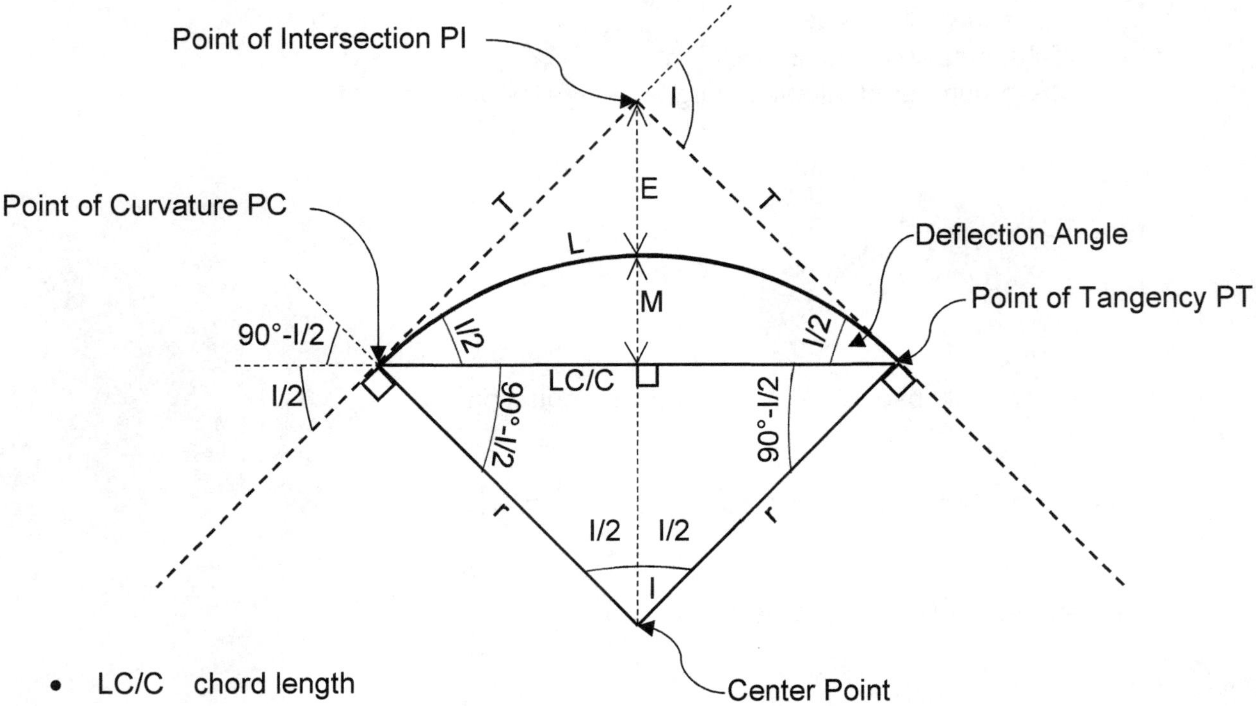

- LC/C chord length
- L arc length
- E external length
- M middle ordinate
- D degree of curve
- I intersection angle = delta angle = central angle
- r radius
- T tangent length
- TD tangent distance: Distance from PC to a point of intersection with a line perpendicular to PT (see image).
- TO tangent offset: Distance of line from PT to the perpendicular intersection with the tangent distance line (see image).

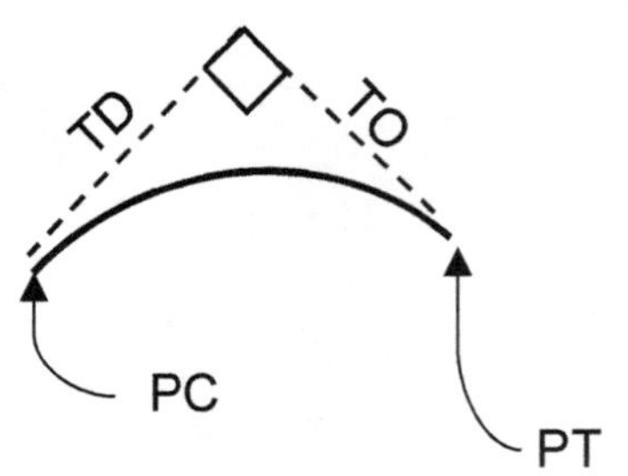

Horizontal curve equations

The use of given elements of the horizontal curve will assist in solving for unknown elements of the horizontal curve. The equations below will allow for the use of these known elements to solve for the unknown elements. Note: There will be times when secondary or even tertiary steps are required to solve for an unknown when given certain elements. An example would be if asked to solve for M and the givens are LC and I, then r will need to be found first in order to proceed.

$$D = \frac{5729.58}{r} \quad (for\ arc\ definition)$$

$$D = \frac{5729.65}{r} \quad (for\ chord\ definition)$$

$$L = rI\frac{\pi}{180°} = \frac{I}{D}(100)$$

$$LC = 2r \sin\left(\frac{I}{2}\right)$$

$$M = r\left[1 - \cos\left(\frac{I}{2}\right)\right]$$

$$E = r\left[\frac{1}{\cos\left(\frac{I}{2}\right)} - 1\right]$$

$$\mathrm{T} = \mathrm{r}\left(\tan\left(\frac{I}{2}\right)\right)$$

$$\mathrm{TD} = \left(cos\frac{I}{2}\right)LC$$

$$\mathrm{TO} = \left(sin\frac{I}{2}\right)LC$$

Know the relationships of angles

Knowing angles will allow for analysis and provide for interpretation of horizontal curves.

Note: Opposing angles of intersecting lines are equal.

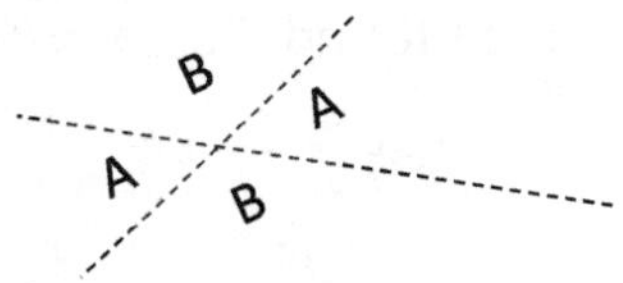

1. Central angle is I, or equal to.
2. Deflection angle is (I/2), or equal to.
3. Radius to tangent is (90°), or equal to.
4. Radius to chord is (90° - I/2), or equal to.
5. Radius to mid-point of chord is (90°), or equal to.
6. Tangent to tangent is (180° - I), or equal to.

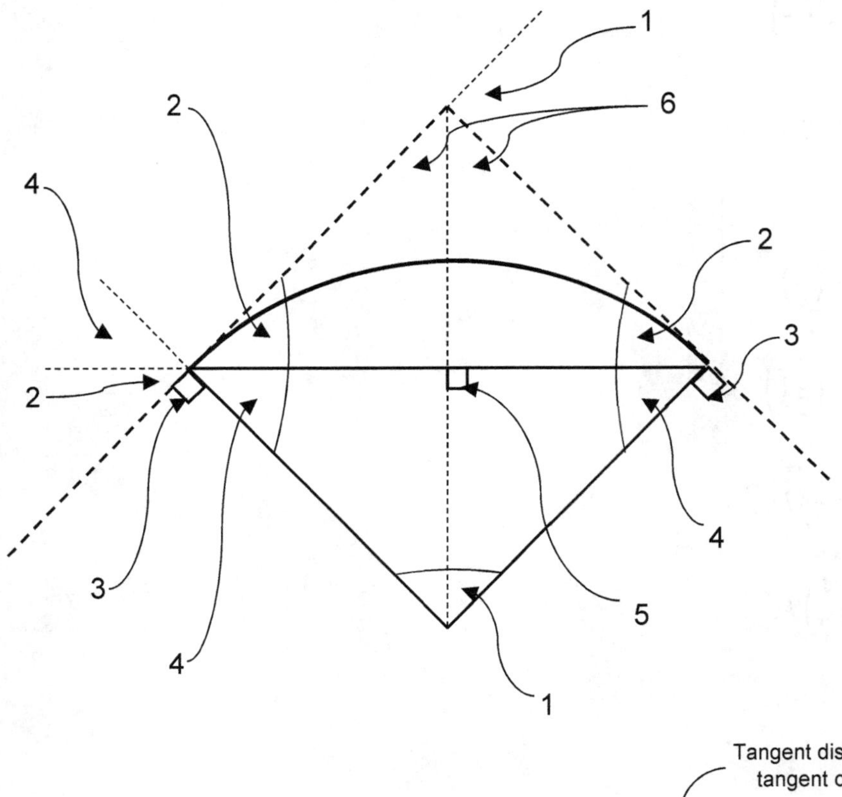

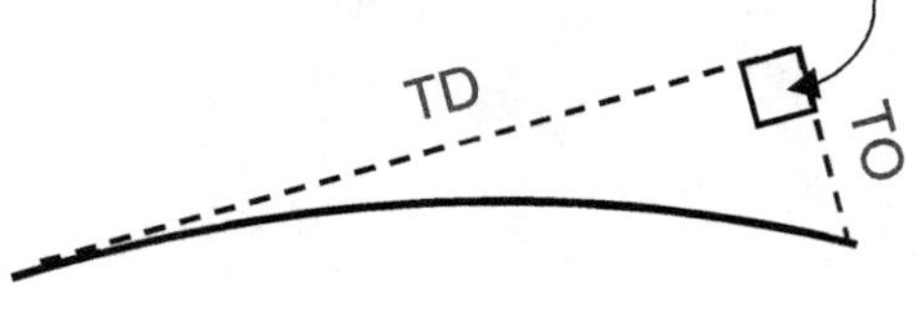

Relationship of degree of curvature D and central angle I

I = D when L = 100' (for the arc definition).
I = D when LC = 100' (for chord definition).

But I and D are proportional even when L or LC do not equal 100'. The concept is the continued curve beyond the defined 100' L or LC maintains the same proportionality. (See the section on curve definitions).

The ratio of the degree of curvature to the arc length of 100' (or chord length if chord definition is considered) equals the ratio of the degree of curvature plus overall angle to 100' plus the extended arc length (or chord length if chord definition is considered).

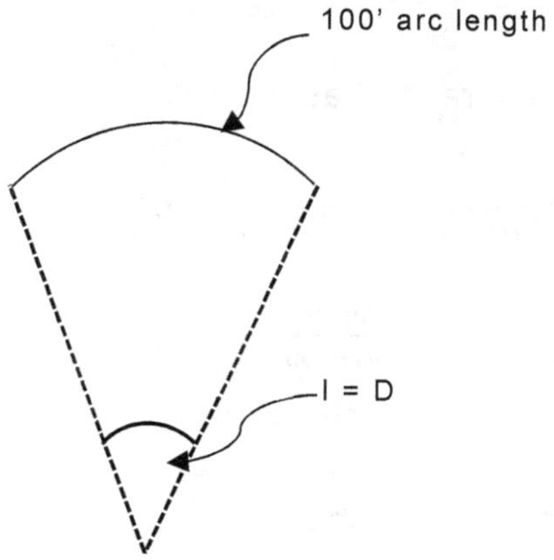

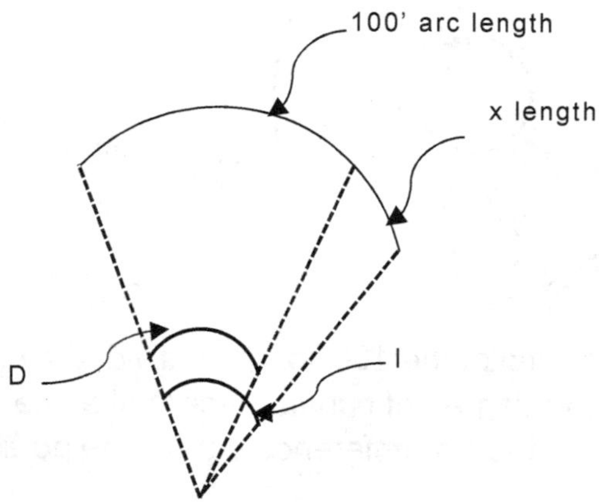

$$\frac{D}{100} = \frac{I}{100 + x}$$

$$(D)(100 + x) = (I)(100)$$

Note: The discussed proportionality is only relevant when pertaining to a continuous curve.

Curve definitions

There are two definitions associated with horizontal curves, the arc definition and the chord definition. The definitions describe the horizontal curve by either arc distance or chord distance.

Arc Definition	Evaluates a curve based on any given portion of the curve where the arc = 100' (most commonly utilized in highway design).
Chord Definition	Evaluates a curve based on any given portion of the curve where the chord = 100' (most commonly utilized in railroad design).

In both definitions the radius length and the degree of curvature are inversely proportional, meaning the larger the degree of curvature the smaller the radius, and the smaller the degree of curvature the larger the radius.

Arc definition

The arc definition is expressed as a ratio:

$$\frac{degree\ of\ curvature}{360°} = \frac{100}{circumference\ (2\pi r)} \quad \text{or} \quad \frac{D}{360°} = \frac{100}{C}$$

Likewise; $$\frac{central\ angle}{full\ circle} = \frac{arc\ length}{circumference} \quad \text{or} \quad \frac{I}{360°} = \frac{L}{2\pi r}$$

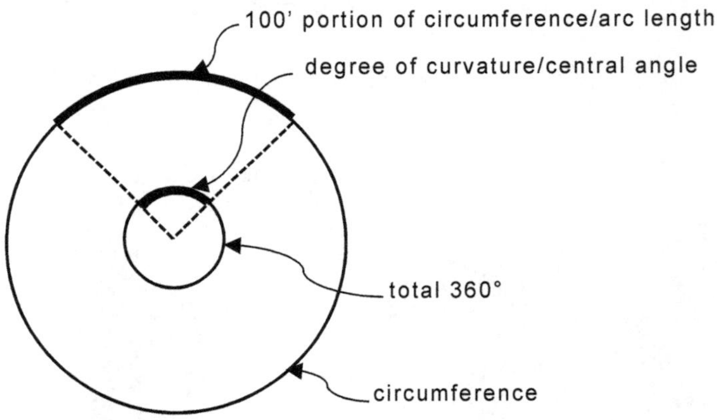

The concept is the ratio of the 100' long arc divided by the circumference will equal the ratio of the degree of curvature/central angle divided by the total 360° circle. The portion of the circumference equals the portion of the 360° circle.

$$\frac{D}{360°} = \frac{100}{2\pi r}$$

$$D \times 2\pi r = 36000$$

$$D \times r = 5729.58$$

$$D = \frac{5729.58}{r} \quad \text{or} \quad r = \frac{5729.58}{D}$$

Chord definition

The chord definition is expressed as a ratio

$$sin(\frac{degree\ of\ curvature}{2}) = \frac{50}{r} \quad \text{or} \quad r = \frac{50}{\sin(\frac{D}{2})}$$

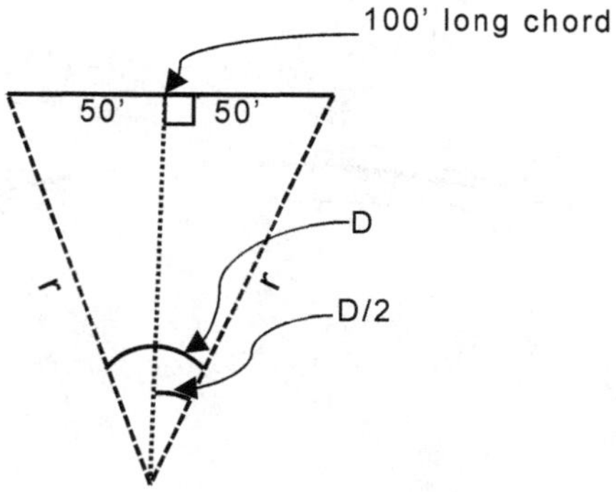

If D = 1°, then

$$sin(\frac{D}{2}) = \frac{50}{r}$$

$$sin(\frac{1}{2}) = \frac{50}{r}$$

$$0.00872653 = \frac{50}{r}$$

$$0.00872653r = 50$$

$$r = 5729.65$$

Thus if D is 1°, then the radius is 5729.65'. Note: When using the chord definition as opposed to the arc definition of a curve, the radius length value will be a slightly larger length than the radius length for the arc definition, which was 5729.58' (as determined above in the arc definition section).

Example 1

Given: r = 110'
I = 22°

Unknown: TD and TO

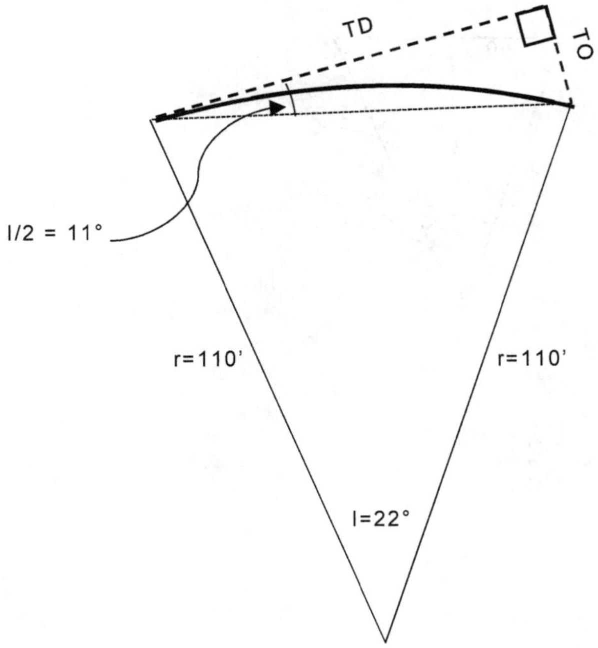

If I = 22°, As determined in the relationships of angle section above, the angle from the chord to the tangents is half of I (in this case 11°).

Next, the LC needs to be determined to utilize the TO and TD equations

$$LC = 2r \sin\left(\frac{I}{2}\right)$$

$$LC = 2(110) \sin\left(\frac{22°}{2}\right)$$

$$LC = 41.9779$$

Finally, solve for TD and TO (using right triangle equations)

$$\text{TD} = \left(cos\frac{I}{2}\right) LC \qquad \text{TO} = \left(sin\frac{I}{2}\right) LC$$

$$\text{TD} = \left(cos\frac{22°}{2}\right) 41.9779 \qquad \text{TO} = \left(sin\frac{22°}{2}\right) 41.9779$$

$$\text{TD} = 41.2066 \qquad \text{TO} = 8.0098$$

Example 2

Given: r = 700'
I = 60°
PC at station 13+20

Unknown: LC at station 14+50
Deflection angle

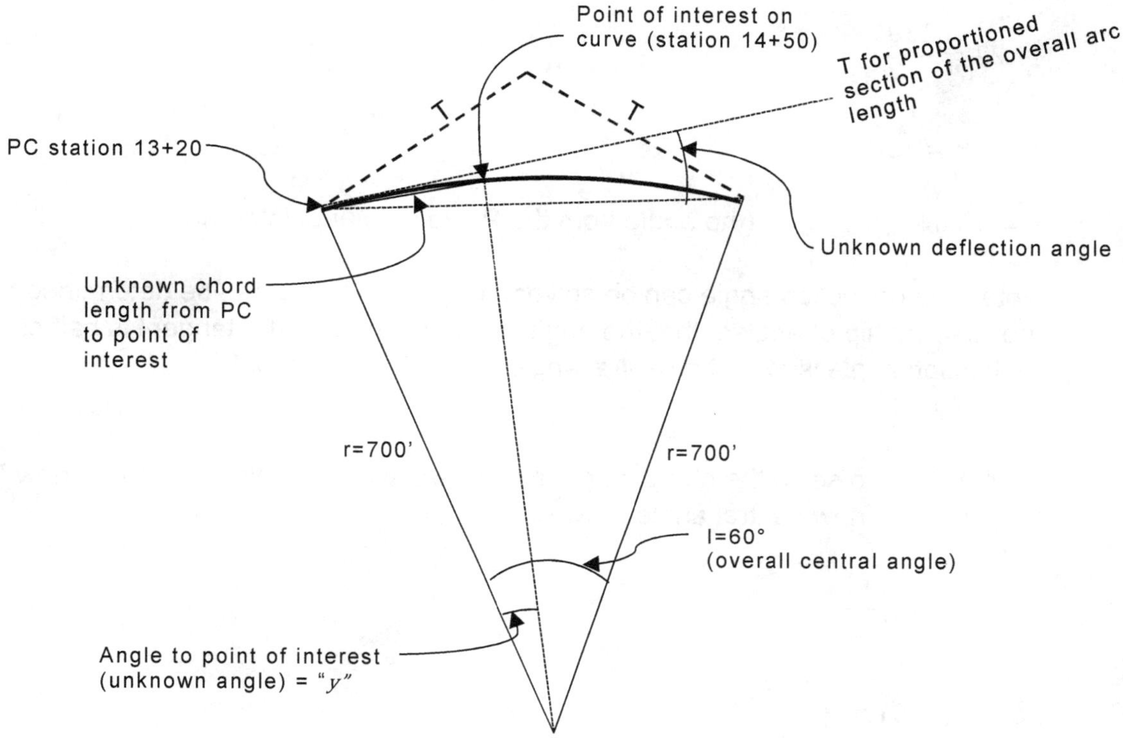

Step 1: The distance from the PC to the point of interest (station 14+50) on the curve is 130'.
(14+50 - 13+20 = 130)

Step 2: Find the overall arc length. This will be necessary to establish the proportionality between the overall arc length L in relation to the overall central angle I, which can then be used to solve for the angle associated with the point of interest y. As discussed in the "relationships between D and I" section, a curve's arc length is proportional to its angle throughout the continuous curve.

$$L = rI\frac{\pi}{180}$$

$$L = 700(60°)\frac{\pi}{180°}$$

$$L = 733.038'$$

Step 3: Set up the ratio.

$$\frac{L}{I} = \frac{arc\ length\ to\ point\ of\ interest}{y}$$

$$\frac{733.038'}{60°} = \frac{130}{y}$$

$$12.22 = \frac{130}{y}$$

$$12.22y = 130$$

$y = 10.64°$ (the angle from the PC to a point of interest)

Note: The deflection angle can be solved. If y = 10.64°, it is can be determined based on the relationship of angles, that the angle from the chord to the tangent is half of y (deflection angle is ½ of the central angle). ½ of 10.64° = 5.32°.

Step 4: Now to solve the chord length associated with the point of interest (now LC), y becomes the new central angle (now I).

$$LC = 2r \sin\left(\frac{I}{2}\right)$$

$$LC = 2(700) \sin\left(\frac{10.64}{2}\right)$$

$LC = 129.81'$ (Note: The chord length is slightly smaller than the arc length of 130', this is correct.)

VERTICAL CURVES

A major aspect of survey is the vertical curve and solving for the different elements of a vertical curve. This section outlines the procedures for solving various elements of vertical curves.

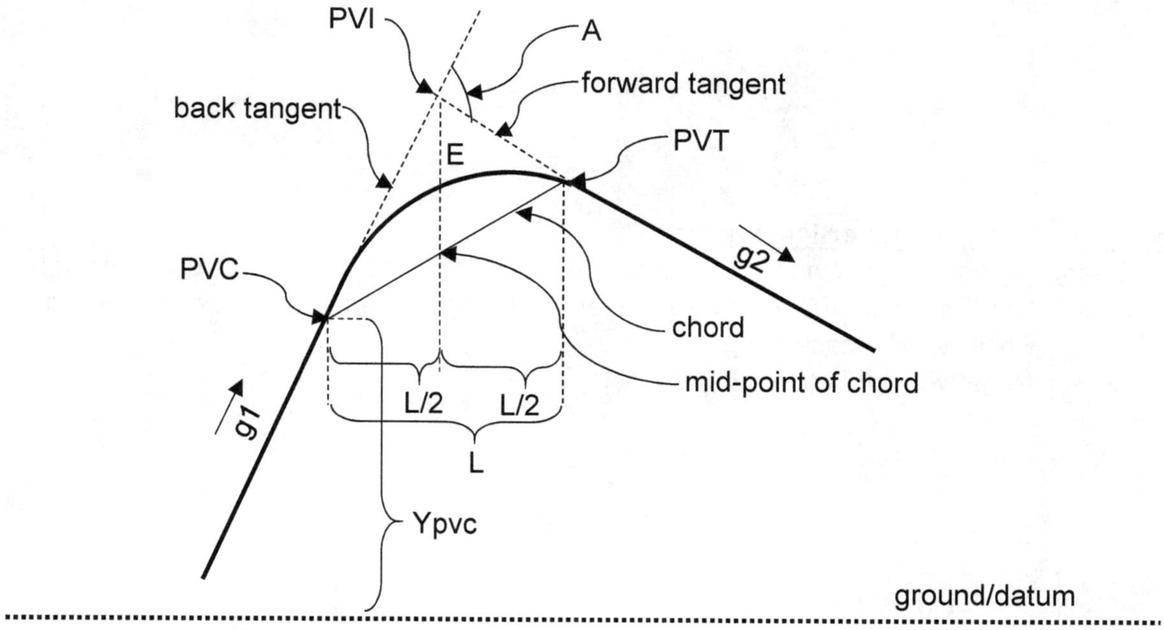

- g_1 grade of back tangent (remember to convert to decimal percent as applicable)
- g_2 grade of forward tangent (remember to convert to decimal percent as applicable)
- PVC point of vertical curve (starting point)
- PVI point of intersection
- PVT point of vertical tangency (ending point)
- L length of curve (horizontal distance) PVI cuts in half when curve is symmetrical
- E tangent offset (vertical distance)
- A change in grade
- Y_{PVC} elevation of PVC
- X distance of point on curve from PVC or PVT (horizontal distance)
- Y change in elevation of point on curve to a tangent line (vertical distance)

Vertical curve types

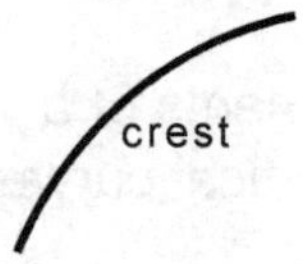

negative change in grade

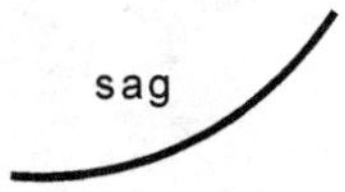

positive change in grade

Aspects and elements of vertical curves

Change in grade = **A**: Change in grade from one tangent to the other is determined by subtracting the grade of the forward tangent g_2 from the grade of the back tangent g_1.

Positive (+) if ascending

Negative (-) if descending

$A = g_2 - g_1$

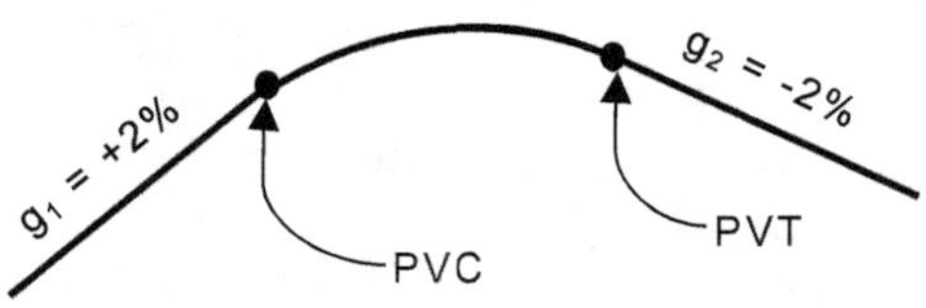

Example:

$A = g_2 - g_1$

$A = -0.02 - 0.02$

$A = \mathbf{-0.04} \approx \mathbf{-4\%}$

Note: When dealing with grades and percent grade DO NOT forget to convert all "percentages" to decimal (e.g., 2%=0.02, 4%=0.04, 1.25%=0.0125).

$\% \div 100 = decimal\ percent$

$decimal\ percent \times 100 = \%$

Slope = **m**: Percent of grade change for a single tangent is determined by the change in y value (known as the RISE) divided by the change in x value (known as the RUN).

$$\frac{RISE}{RUN} = \frac{\Delta y}{\Delta x} = slope$$

$$\frac{\Delta y}{\Delta x} \times 100 = \%\ grade$$

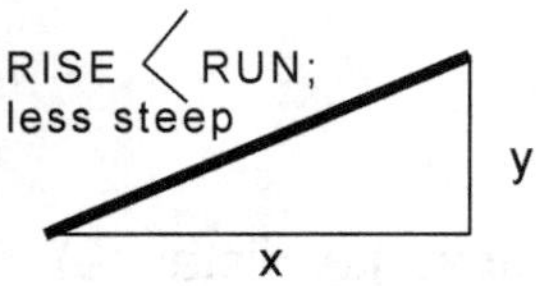

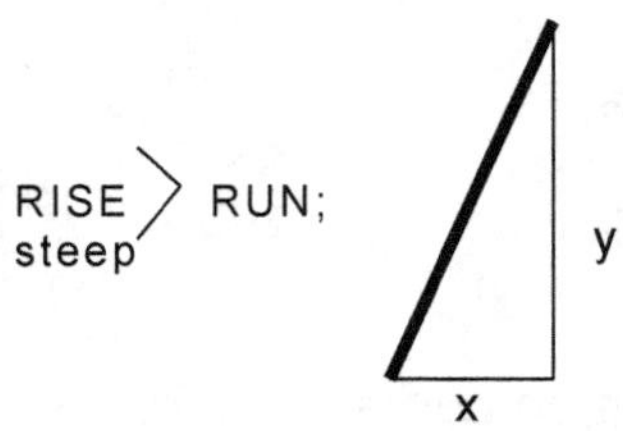

Rate of change of grade = **r**: The rate of change of grade is the change of grade A over the total horizontal distance L of the vertical curve.

$$r = \frac{g_2 - g_1}{L} = \frac{A}{L}$$

Parabolic constant = **a**: For a vertical curve to be parabolic, it must:

1- Have a constant rate of change of slope; smooth flow.
2- Have equal curve tangents; the horizontal length between PVC and PVI equals the horizontal length between PVI and PVT.

$$a = \frac{g_2 - g_1}{2L} = \frac{A}{2L} = \frac{r}{2}$$

Tangent offset at PVI = **E**:

$$E = a\left(\frac{L}{2}\right)^2$$

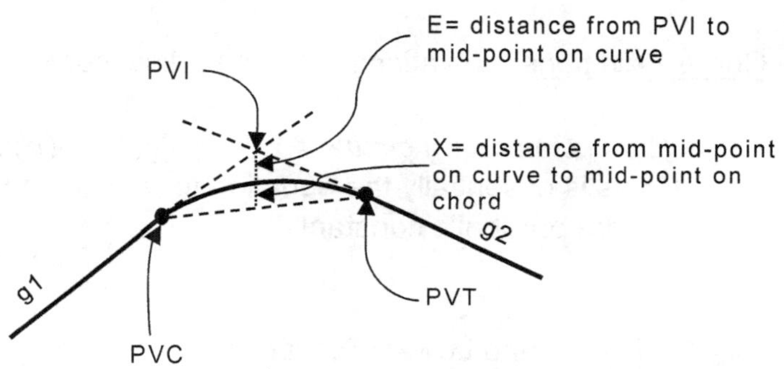

Note: E equals the distance from the chord of curve X if curve is symmetrical.

The tangent offset at any given point takes into account the following proportioning:

$$\frac{tangent\ offset\ at\ a\ given\ point}{\substack{squared\ distance\ from\ PVC\ for\ back\ tangent \\ (or\ PVT\ for\ forward\ tangent)}} = \frac{tangent\ offset\ at\ PVI{=}E}{\substack{total\ squared\ distance\ of\ back\ tangent \\ (or\ forward\ tangent)}}$$

This is stating the tangent offset distance along the tangent at any given point is proportionally derived from knowing the tangent offset distance at the PVI (known as “E”) and the total distance of said tangent.

Tangent elevations: Elevations along the tangents.

$$Elevation\ along\ \textbf{back tangent}\ (PVC\ to\ PVI) = Y_{PVC} + (g_1)x$$

This is essentially stating, an elevation along the back tangent (from PVC to PVI) is determined by the elevation of the PVC plus the tangent grade for the back tangent times a distance along said back tangent from the PVC x.

$$Elevation\ along\ \textbf{forward tangent}\ (PVT\ to\ PVI) = Y_{PVI} + g_2(x - \frac{L}{2})$$

This is essentially stating, an elevation along the forward tangent (from PVT to PVI) is determined by the elevation of the PVI plus the tangent grade for the forward tangent times a distance along said back tangent from the PVT x minus half the length of the curve. Note: The distance along the tangent needs to be the distance from PVT to PVI, and not PVI to PVT (opposite direction than the back tangent elevations equation).

Curve elevations: Elevations along a vertical curve.

$$Elevation\ along\ curve = Y_{PVC} + (g_1)x + (a)x^2$$

This is essentially the equation for elevations along the tangent with the inclusion of the parabolic constant.

Highest and Lowest Elevations on a curve.

$$highest\ or\ lowest\ point\ distance\ on\ a\ curve\ from\ PVC = \frac{(L)|g_1|}{|g_1 - g_2|}$$

A tangent line of a vertical curve at the highest or lowest point on the curve will always be a horizontal line.

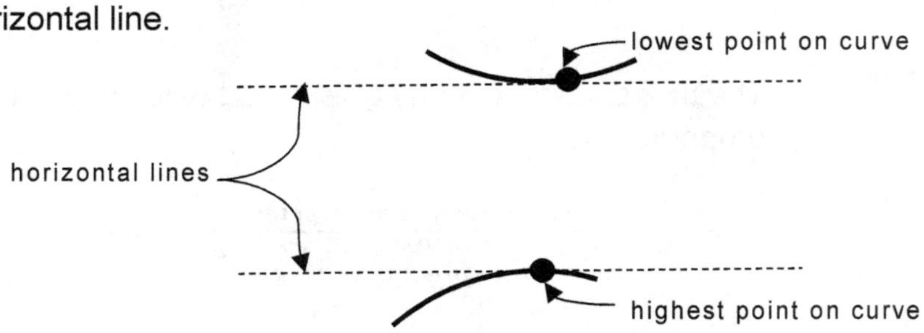

Horizontal distances are NOT slope/arc distances

Horizontal distances are considered during vertical curve calculations, not slope or arc distances.

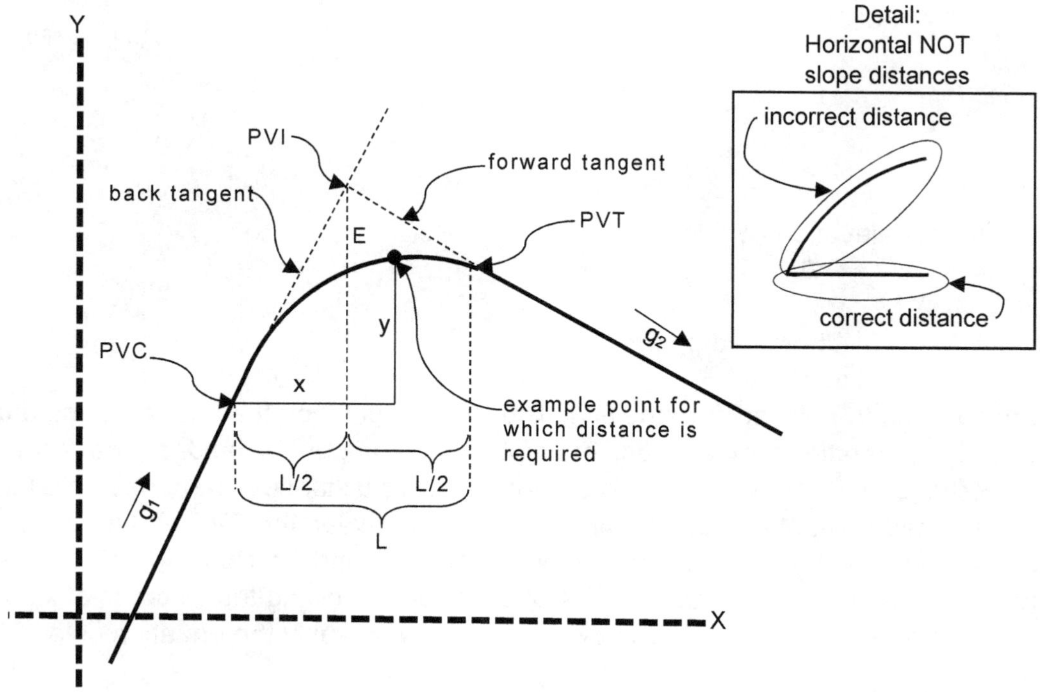

The distance for L above is a horizontal distance and does not consider the distance along the curve. L is simply the change in x value on a grid (See detail).

The E value is the change in y value on a grid.

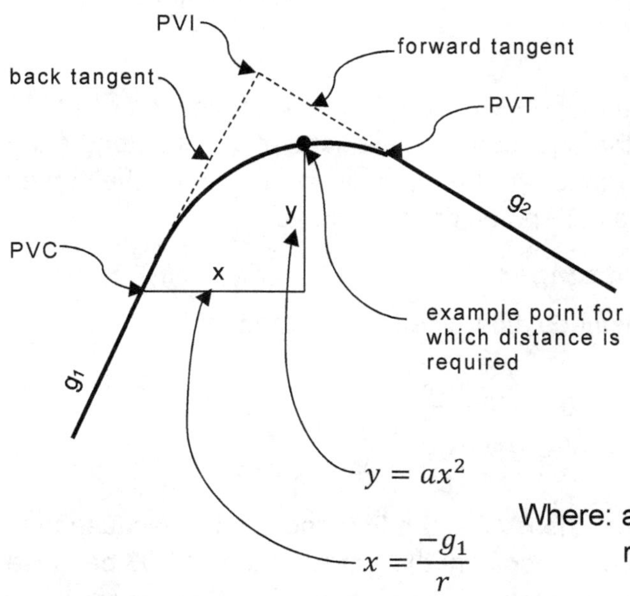

Where: a=parabolic constant

r= rate of change of grade

Example 1

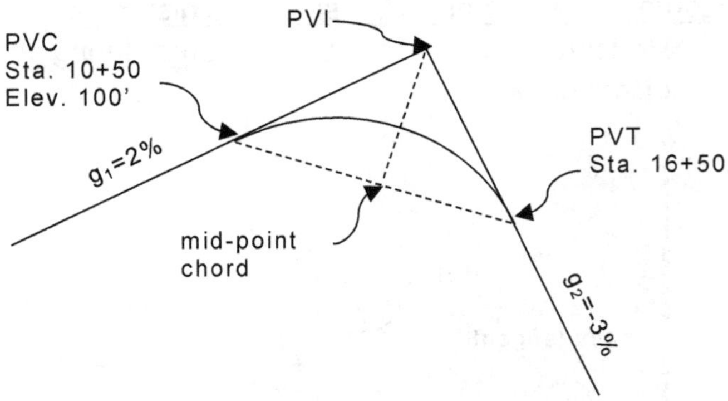

Unknowns elevations of:

PVI

PVT

Chord mid-point

Step 1: Find the elevation of the PVI. The PVI is a point on the back tangent, thus the use of the equation for elevations of point along a tangent is needed (specifically, the back tangent). Keeping in mind 1) all vertical curve distances are horizontal 2) the PVI and the mid-point of chord are half the distance between the PVC and the PVT. It is known then, the mid-point station between the PVC and the PVT is 13+50. Thus, L= 600 (which is 10+50 – 16+50) and (L/2) = 600/2= 300'. Knowing the distance of x as the distance from PVC to PVI, it is known x is 300'. Now apply the equation below.

$$Elevation\ along\ \mathbf{back\ tangent}\ (PVC\ to\ PVI) = Y_{PVC} + (g_1)x$$

Where: x is the distance along the back tangent from the PVC

$$Elevation\ of\ PVI = 100' + (0.02)300'$$

Do not forget to convert the percent grade to decimal. Additionally, the elevation of the PVC was provided.

$$Elevation\ of\ PVI = \mathbf{106'}$$

Step 2: Find the elevation of the PVT. The PVT is a termination point of the forward tangent, thus the use of the equation for elevations of a point along a tangent is needed (specifically, the forward tangent). The equation for a point on the curve could also be used, because the PVT is the end point of the curve.

$$Elevation\ along\ \mathbf{forward\ tangent}\ (PVT\ to\ PVI) = Y_{PVI} + g_2(x - \frac{L}{2})$$

Where: x is the distance along the forward tangent from the PVT

$$Elevation\ of\ PVT = 106' + (0.03)(0 - \frac{600}{2})$$

$$Elevation\ of\ PVT = 106' - 9' = \mathbf{97'}$$

Note: x is the distance from the PVT to the PVI and is zero because the distance from the PVT to the PVT is zero. In addition, the grade used is +0.03 because we are going <u>up</u> a negative grade. g_2 is negative (-0.03), but the direction is from the PVT to PVI so the grade will be positive (+0.03).

Step 3: Find the elevation of the mid-point of chord. If the chord is a straight line between the PVC and the PVT then the mid-point of this line (the chord) should be half way between the PVC and PVT both in terms of its x value and y value. The y valve is associated with elevations in regards to vertical curves. A simple average of the PVC and PVT elevations will provide the mid-point elevation.

PVC elevation = 100'
PVT elevation = 97'

100' + 97' = 197'

197' ÷ 2 = 98.5'

The elevation of the mid-point is **98.5'**

Example 2

Using the same vertical curve as depicted in the previous example and using the givens and determined values, what is the longest tangent offset distance?

The "longest tangent offset distance" will be the distance from the PVI to the curve; essentially the E value. The equation for E is:

$$E = a\left(\frac{L}{2}\right)^2$$

Step 1: solve for a (parabolic constant)

$$a = \frac{g_2 - g_1}{2L}$$

$$a = \frac{-0.03 - (0.02)}{2(600)}$$

$$a = 0.00004166666$$

Step 2: solve for E

$$E = a\left(\frac{L}{2}\right)^2$$

$$E = 0.00004166666\left(\frac{600}{2}\right)^2$$

$$E = \mathbf{3.75'}$$

INTERSECTING LINE:
CENTER CORNER RESTORATION

In sectionalized lands typically (though exceptions do exist) the center of a section or center of quarter section are established by the intersection of lines. The intersection of lines produces a point of intersection. The methods in this section outline the process of identifying the coordinates of the intersection point of two intersecting lines.

Slope

The slope of each line needs to be established before the intersection point is determined. Slope is the difference in y values divided by the difference in x values. Slope is known as "m".

$$m = \frac{\Delta y}{\Delta x}$$

Note: It is very important to keep coordinates "aligned" when determining slope.

$$m = \frac{y_1 - y_2}{x_1 - x_2} \; or \; m = \frac{y_2 - y_1}{x_2 - x_1}$$

Note: Negative slope, the line moves downward left to right. ↘
Positive slope, the line moves upwards left to right. ↗

Y-intercept

Next, the y-intercept needs to be determined for each line. Y-intercept is known as "b". Y-intercept is the y value when x equals zero; where the line cross the y-axis.

$b = y - mx$

Note: Slope intercept form is $y = mx + b$.

Steps in determining intersection point of two straight lines

1) Find slope; m.
2) Find y-intercept; b.
3) Find x value of the intersection point; $mx + b = mx + b$ (slope intercept form equation), by setting each line's slope intercept form equation equal to one another.
4) Find y value of the intersection point by inserting the x value found into either line's slope intercept form equation.

Example

Line 1 coordinates:
(1 , 5)
(15 , 10)

Line 2 coordinates:
(10 , 5)
(4 , 11)

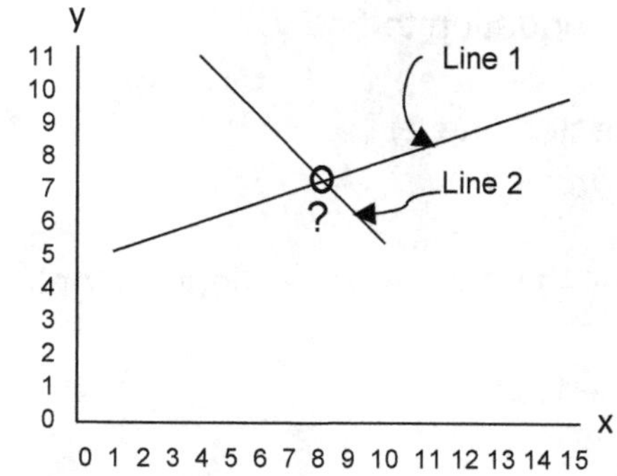

Step 1: Determine the slope for each line.

Line 1 $m = \frac{\Delta y}{\Delta x}$

$m = \frac{5-10}{1-15} = \mathbf{0.357}$

Check if it matches the line orientation as graphed; Yes ↗

Line 2 $m = \frac{\Delta y}{\Delta x}$

$m = \frac{5-11}{10-4} = \mathbf{-1}$

Check if it matches the line orientation as graphed; Yes ↘

Step 2: Determine the y-intercept for each line. Using any point from each line as your x and y values.

Line 1 $b = 5 - (0.357)1$

$b = \mathbf{4.643}$

Line 2 $b = 5 - (-1)10$

$b = \mathbf{15}$

Step 3: Determine the x value of intersection by setting each line's slope intercept form equation equal to one another.

Line1 $(mx + b)$ = Line 2 $(mx + b)$

$(0.357)x + 4.643 = (-1)x + 15$

$0.357x + 4.643 = -1x + 15$

$0.357x + 4.643 = -1x + 15$

$\underline{\quad -15 \qquad\qquad - 15}$

$0.357x - 10.357 = -1x$

$\underline{-0.357x \qquad\qquad - 0.357x}$

$-10.357 = -1.357x$

$\mathbf{7.632} = x$

Step 4: Insert the newly found x value from step three into either line's slope intercept form equation to find y.

(selected line 2)
$y = mx + b$

$y = (-1)7.632 + 15$ slope m for line 2 was (-1) and the y-intercept b was 15.

$y = -7.632 + 15$

$y = \mathbf{7.368}$

The coordinate for the intersection point of the two straight lines is **7.632, 7.368**

PHOTOGRAMMETRY

Elements of photogrammetry

Scale: The proportional relationship between the actual ground and the imagery of the ground.

$$scale = \frac{distances\ in\ image\ or\ photography}{distances\ on\ ground} = \frac{ab}{AB} = \frac{f}{H}$$

Where: f is the focal length (the distance from sensor to lens)
H is the average height/altitude

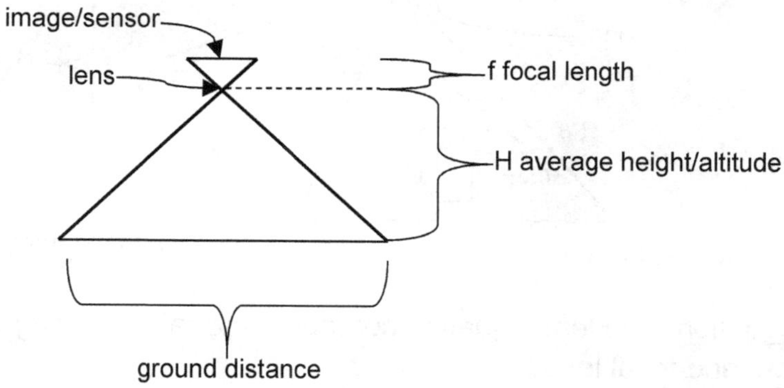

Note: It is assumed the imagery/photograph is perfectly vertical and not angled in any way.

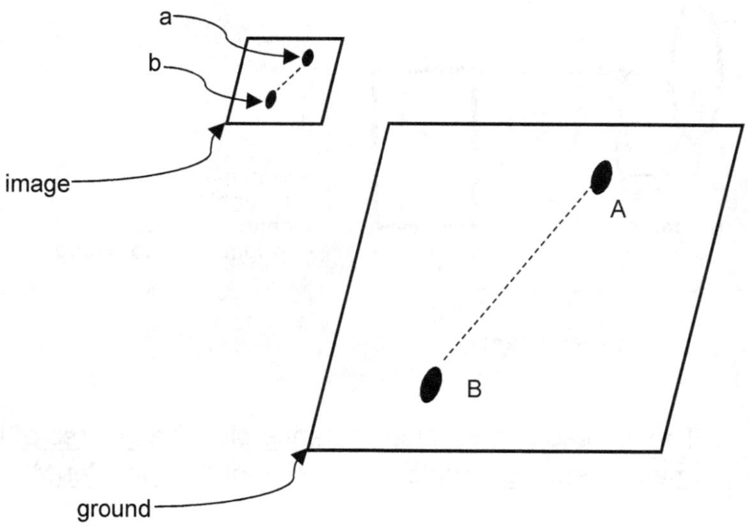

Relief displacement: If topographic features are captured with imagery the further they are from the center of the image the more distorted (leaning) they will appear. Additionally, the taller the feature the more distortion. This displacement increases radially outward from the image center point known as the principal point.

$$d = \frac{rh}{H}$$

Where: d is the relief displacement
r is the radial distance from the image center/principal point
h is the feature height
H is the height/altitude

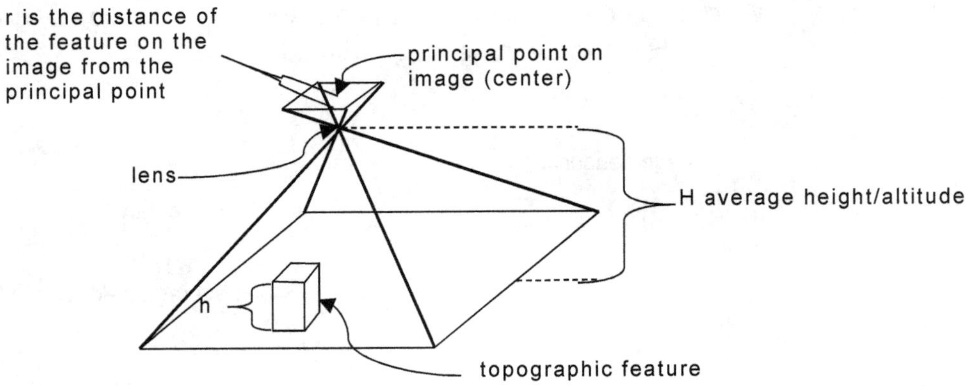

Lens equation: The lens equation compares the ratios among object distance, image distance, and focal length.

Object distance: Distance between actual object and lens o.
Image distance: Distance between image and lens i.
Focal length: Distance between focal point and lens f.

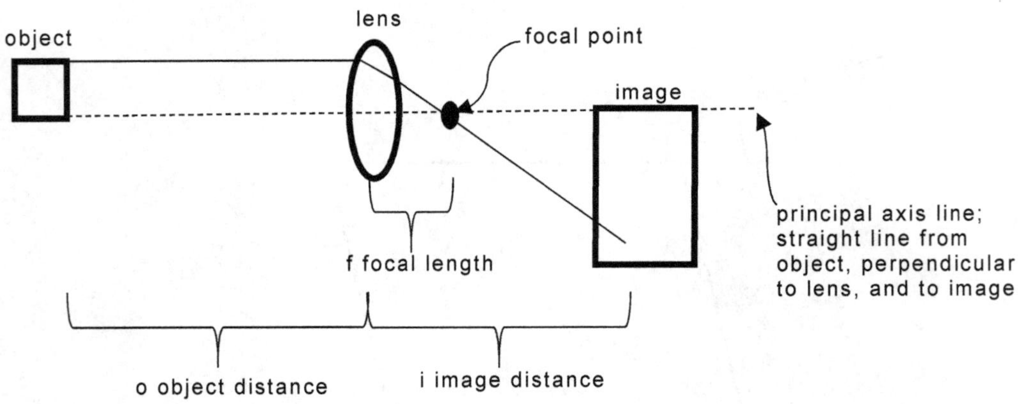

$$\frac{1}{o} + \frac{1}{i} = \frac{1}{f}$$

The inverse of the object distance plus the inverse of the image distance equals the inverse of the focal length

Parallax equations: The apparent shift in an object's position from different viewpoints. An example of this is the shifting of an object's position from one eye to the other (looking at thumb with left and right eyes individually). The concept behind this is the closer the object is to the viewpoint (e.g., camera, eye, etc.) the greater the shift will be if the viewpoint is shifted. The further away the object; the apparent shift is smaller. The apparent shift is in the same direction as the motion of the viewpoints. In the diagram below, it is assumed the direction of motion is along the x-axis; thus, no parallax will be observed along the y-axis.

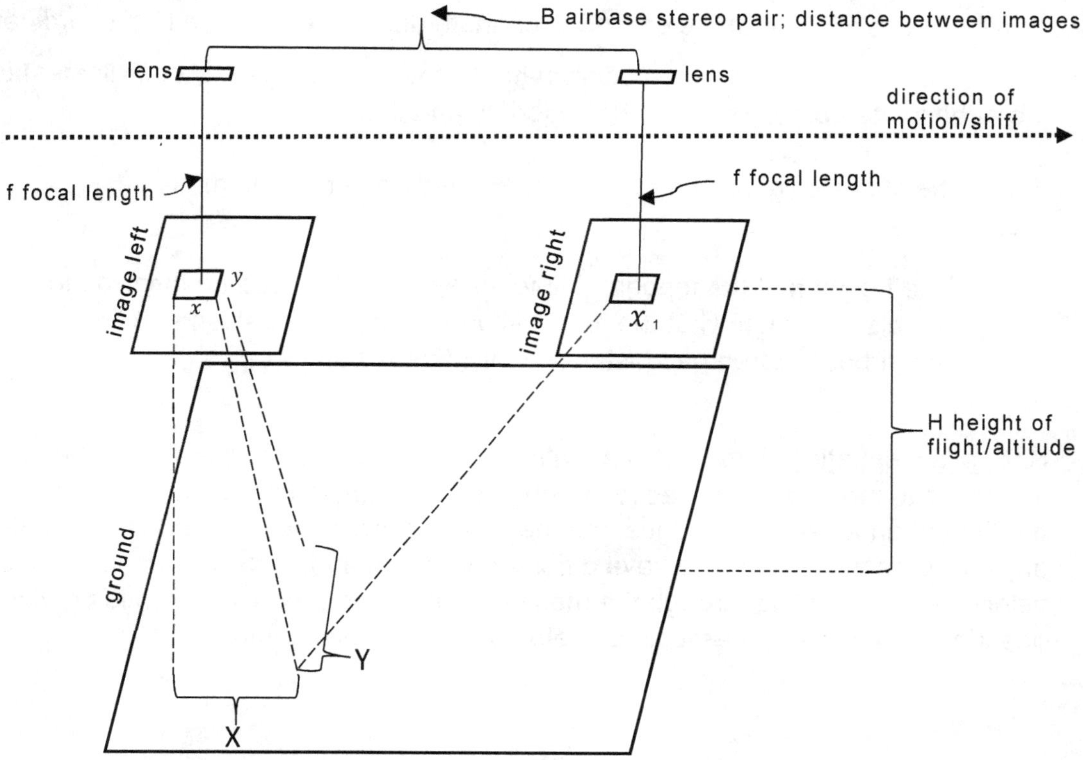

$parallax\ (p) = x - x_1$

The distance between the coordinates observed from the left image minus the coordinates observed from the right image.

(Example: if the distance from a light pole base to a tree stump in image one is 5" and the distance from a same light pole base to a same tree stump in image two is 5.5", then the parallax would be $(x - x_1)$, or (5.5" – 5") = 0.5"

Now that parallax is found, then the application of parallax to find the X and Y distances can be performed:

$X = \frac{xB}{p}$ The X value of the ground coordinate difference (change in X coordinates) equals the x value of the image coordinate difference (change in x coordinates) times the airbase stereo pair distance B divided by parallax.

$Y = \frac{yB}{p}$ The Y value of the ground coordinate difference (change in Y coordinates) equals the y value of the image coordinate difference (change in y coordinates) times the airbase stereo pair distance B divided by parallax.

Finally, height calculations can be derived based on photogrammetry.

$h = H - \frac{f(B)}{p}$

The height h of the topographic feature being observed on the ground equals the height H of the flight/altitude minus the focal length f times the airbase stereo pair distance B divided by parallax p.

<u>Snell's law equation</u>: The Snell's equation provides for the comparison of the angle of incidence to the angle of refraction through different mediums. When light enters a medium at an angle of incidence, the medium will divert the light and create a different angle of refraction. Mediums have different densities and, in turn, light has a different velocity while traveling through the medium. A medium's index value n is provided to describe its density and associated velocity of light passing through it.

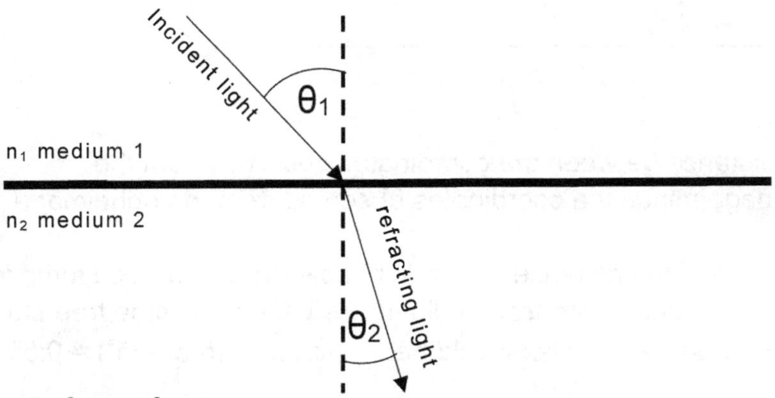

$n = index\ value$

$\theta = angle$

Snell's law equation states:

$n_1(sin\theta_1) = n_2(sin\theta_2)$

The index value of medium 1 times the Sine of the incident angle through medium 1 will equal the index value of medium 2 times the Sine of the refraction angle traveling through medium 2.

Curvature and refraction equation: The curvature of the Earth creates a deviation of the line of sight from a perfect horizontal plane; the further the distance the larger the error. Additionally, the atmosphere (medium in which the light is traveling) creates a bending of the light downward known as refraction. This causes objects to appear higher than they actually are. Temperature and pressure of the atmosphere affect the intensity of bending. While surveying, the observance through a total station will have both curvature and refraction acting on the line of sight.

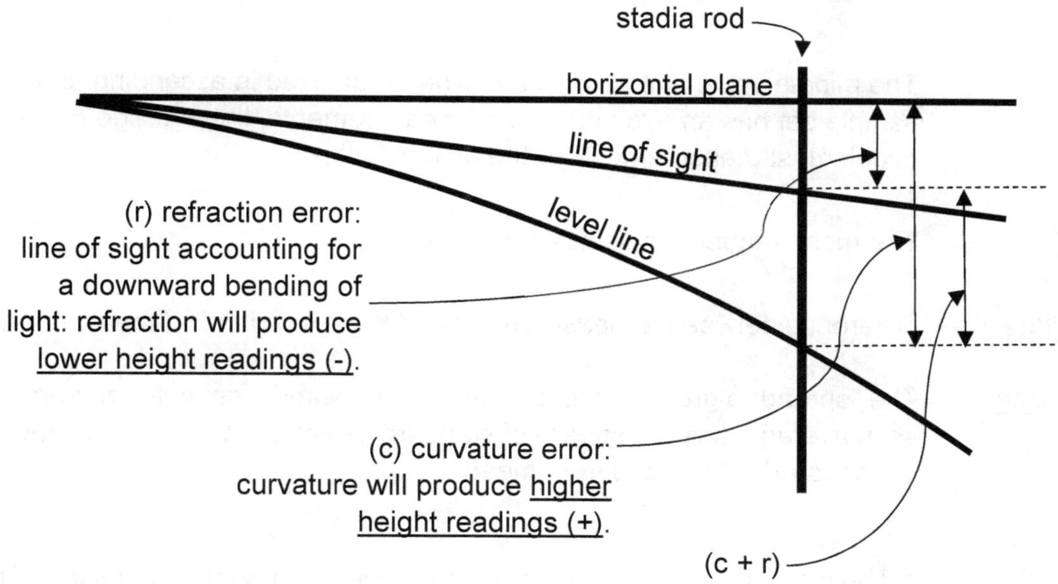

Note: The images above depict the acts of curvature and refraction.
Note: The error or refraction is 1/7 of the error due to curvature.

c = Correction for curvature is $(0.0206)distance^2$

r = Correction for refraction is $\frac{(0.0206)distance^2}{7}$

When correcting for curvature and refraction, the following equation is used:

$(c + r) = 0.0206M^2$

Where: c is curvature in feet
r is refraction in feet
M is distance in thousands of feet

STATISTICS AND PROBABILITY

Statistics is the science of assuming smaller or proportional determinations from the analysis of larger sets of data. Statistics will be utilized for the purpose in various ways such as least squares adjusting of best fit lines and other applications.

Mean — Average $\bar{x}$ of a sample set $\frac{x_{1+}x_n}{n}$.

Median — The middle value of a sample set, when organized in ascending order. If the sample set has an even number of measurements, the average of the two middlemost measurements is the median value.

Mode — The most frequent value of a sample set.

Residual — Difference between a measurement and the average of all measurements.

Variance — The "spread" a group of measurements in a sample set extends from the sample set's average. It is the square of standard deviation, and is the average (n-1) of the squared differences from mean.

Standard Deviation — σ The quantity found to determine the range or amount of deviation of a group of measurements. In addition, it is the square root of the variance.

Variance

Variance is the average of the squared differences of each measurement from mean. The variance allows for the determination of the standard deviation. The square root of variance is the standard deviation.

Standard deviation

Standard Deviation is the square root of variance (square root of the average of the squared differences of each measurement from mean). The standard deviation reveals the limits on either side of the mean for the first standard deviation and subsequent standard deviations (refer to the section on the "68-95-99.7 rule" below for further explanation).

$$\sigma = \sqrt{\frac{\Sigma(x_i-\bar{x})^2}{n-1}} = \sqrt{\Sigma\frac{v^2}{n-1}}$$

Where:

σ	the standard deviation of measurements
v	the variance of measurements
$\bar{x}$	the mean of the sample set of measurements
x_i	the designation of an individual measurement
n	the number of measurements in the sample set
Σv^2	the sum of the squares or the residuals

Example

Given 6 individual measurements as follows:

10, 6, 8, 7, 8, and 9

Step 1: Calculate the average: $\frac{10+6+8+7+8+9}{6} = \mathbf{8}$

Step 2: Find the difference/residual for each measurement from the average:
10-8=**2**, 6-8=**-2**, 8-8=**0**, 7-8=**-1**, 8-8=**0**, and 9-8=**1**

Step 3: Square each the found residuals:
2^2=**4**, -2^2=**4**, 0^2=**0**, -1^2=**1**, 0^2=**0**, 1^2=**1**

Step 4: Calculate the variance (average of the squared residuals; (n-1)): $\frac{4+4+0+1+0+1}{6-1} = \mathbf{2}$

Step 5: Calculate the standard deviation (the square root of the variance):
$\sqrt{2} = \mathbf{1.414}$ is the standard deviation.

One standard deviation is 1.41 units either side of the mean. See the image below showing the plotted points, the mean line, and the standard deviation lines either side of the mean line.

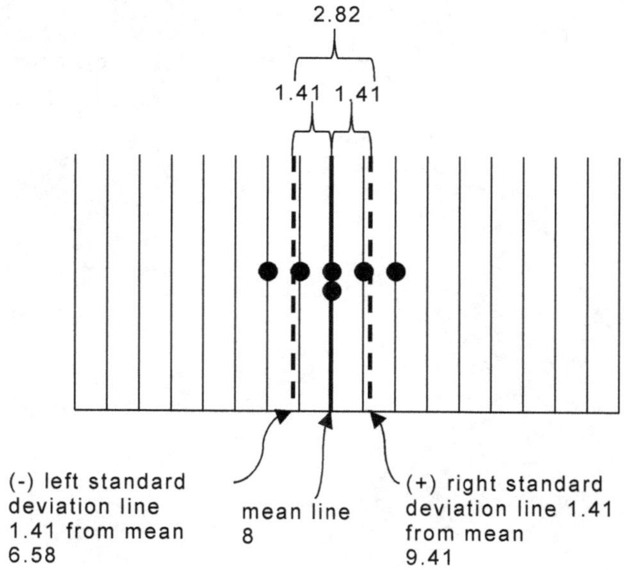

The percentages associated with the standard deviations follow the "68-95-99.7 rule". This means 68% of measurements will fall within the first standard deviation, 95% of measurements will fall within the first and second standard deviations, and 99.7% of measurements will fall within the first, second, and third standard deviations. The depiction below is for normal distributions.

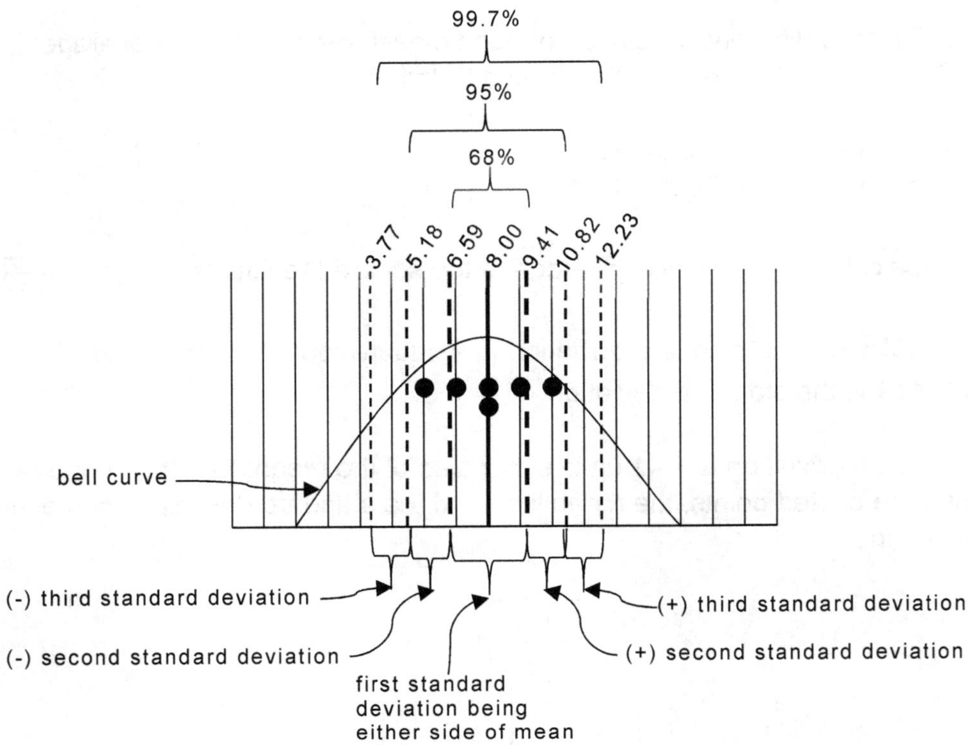

ERRORS, PROPAGATION OF ERROR, UNCERTAINTY, AND TOLERANCE

With measuring, there will be errors that occur. Nothing is ever perfectly measured, but the goal is to understand the types of error, identify the error, and to finally impose a method to minimize the error.

Types of Error

There are three types of error:

Systematic errors - the type of error that is identifiable and thus can be corrected for. Systematic errors typically produce a fluctuation of measurements in a consistent or proportional manner (e.g., all too high or all too low). Systematic errors are further divided into four types:

Instrumental errors - errors caused by poorly calibrated instruments being used to measure. An example could be incorrect prism constant, incorrect rod height, etc.

Observational errors - errors that occur due to the different interpretations of the observer(s), different perspectives of the observer(s), observation by different instruments or different instrument positions, etc. An example could be parallax of aerial imagery.

Environmental errors - errors that occur due to environmental conditions or situations such as temperature, humidity, elevations, etc.

Theoretical errors - errors that occur due to simplification, incorrect assumption, or approximation in measurements.

Random errors - the type of error that is unidentifiable and unpredictable and thus cannot be corrected for. Random errors typically produce different fluctuations from one measurement to the next (e.g., first is too high, second is too low, and third is too high). Because random errors are uncontrolled and unbiased they cause a fluctuation in all directions. Due to this fact, random errors tend to eliminate or reduce themselves, but never do so completely. Random error are further handled by propagation of error (see propagation of error section below).

Blunder errors - the type of error that is simply a mistake or carelessness in measurements. An example of a blunder may be a typo made by the observer.

Accuracy vs. precision

Accuracy refers to the how close the measurement is to the true value (small residuals). Accuracy is the agreement with any given measurement to the true value.

Precision is completely separate from accuracy and refers to the relationship of values to each other as opposed to the true value. Precision is the agreement of one measurement to other measurements in the same system. Better the precision the smaller the random errors being introduced into the measurements.

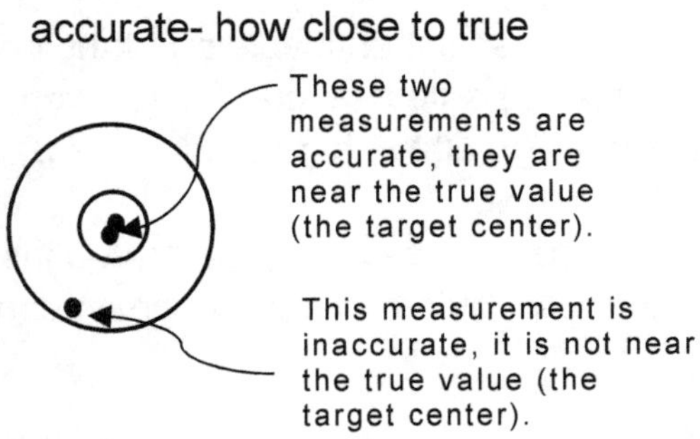

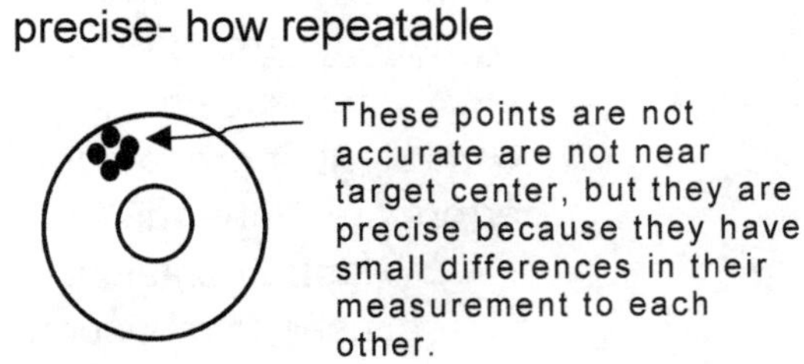

Uncertainty

Uncertainty is the range of possible values within which the true value of a measurement exists. Uncertainty is an expression of the level of doubt about a measurement. Uncertainty is typically written in "+/-" value (e.g., 5.5' +/-0.3'; meaning the true value of the measurement lies within the range of 5.5' **+** 0.3' = **5.8'** and 5.5' **-** 0.3' = **5.2'**). Uncertainty is commonly expressed in terms of standard deviation (see section on statistics and standard deviation for further definition and explanation). If uncertainty is expressed in terms of standard deviation then the range of uncertainty (the +/- value) is the standard deviation value.

The concept of uncertainty is as follows:

measurement = best estimate +/- uncertainty

The collection of repeated measurements will provide the values to produce the standard deviation of the values. Then, any single value from the repeated measurements has an uncertainty equal to the standard deviation.

Positional tolerance or "measurable tolerance"

Tolerance is the pre-determined allowable amount assigned to a measurement can vary from the value it is attempting to measure. It is the permissible range of variation. Uncertainty is derived from statistical analysis and tolerance is typically a set industry standard (this standard may or may not be based on statistical analysis).

Positional accuracy

Positional accuracy is expressed as relative accuracy and as absolute accuracy.

Positional relative accuracy - the measurement of map positions in relation to each other. The comparison of one point/feature on a map to a different point/feature on the same map.

Positional absolute accuracy - the measurement of map positions in relation to the true ground position it represents. The comparison of a mapped point/feature location to the true ground location.

Propagation of error or "random error propagation" or "propagation of uncertainty"

First, all measurements have random error. Second, the definition of an error is the difference between the true value and a measured value attempting to identify said true value. Note that error is unavoidable.

The concept or error is as follows:

error = measurement value – true value

As previously stated, random errors tend to nearly cancel or eliminate themselves, but never completely do so. Thus, there needs to be an understanding of the accumulation and cancellation for random errors introduced into a system. This concept of understanding random error behavior is known as error propagation or random error propagation.

Summation of different errors: when random errors introduced into a system are all different, the following equation becomes applicable:

Where:

$\sigma_{sum} = \sqrt{\sigma_1^2 + \sigma_2^2 + \sigma_3^2 \dots\dots + \sigma_n^2}$

σ= the uncertainty of a measurement

n= sample size

Example:

A row team can row a boat for one mile with the uncertainty of +/- 10 seconds when clocking them. Additionally, the river the row team is going to perform the one mile row has a constant flow rate with an uncertainty of +/- 3 seconds. In this example, there are two different types of uncertainty associated with measuring the time it will take to row one mile (row rate and flow rate). If the formula is overall time to row one mile equals the time to row minus flow rate (if traveling downstream/with the river flow).

time = (row rate) – (downstream river flow rate)

If the row team can row a mile in 45 seconds (+/- 10 seconds of uncertainty) and the river flow rate is 15 seconds (+/- 3 seconds of uncertainty), then the following would apply:

time = (row rate) – (river flow rate)
time = (45) – (15)
time = 30 seconds

However, the level of uncertainty is:

$\sigma_{sum} = \sqrt{\sigma_1^2 + \sigma_2^2 + \sigma_3^2 \ldots\ldots + \sigma_n^2}$

$\sigma_{sum} = \sqrt{10^2_{row\ rate\ uncertainty} + 3^2_{river\ rate\ uncertainty}}$

$\sigma_{sum} = +/-10.44\ seconds$

Finally, the row team can row one mile in this situation in 30 seconds (+/- 10.44 seconds). Therefore, they can do it somewhere between 40.44 seconds and 19.44 seconds.

Errors in series: when random errors introduced into a system are all the same, the following equation becomes applicable.

$\sigma_{series} = \sigma\sqrt{n}$

σ= the uncertainty of a measurement

n= sample size

Example:

Using the example from above, a row team can row a boat for one mile in 45 seconds with the uncertainty of +/- 10 seconds when clocking them. If the row team rows the one mile stretch a total of ten times, then the error in series equation would apply as follows:

First, the row team can row the 10 sessions of the one mile lap in a total of 450 seconds (10 sessions times 45 seconds per lap).

Secondly, the uncertainty can be calculated as follows:

$\sigma_{series} = \sigma\sqrt{n}$

$\sigma_{series} = 10\sqrt{10}$

$\sigma_{series} = +/-31.62$

Finally, the row team can row ten sessions of a one mile lap in this situation in 450 seconds (+/- 31.62 seconds). Therefore, they can do it somewhere between 481.62 seconds and 418.38 seconds.

<u>Averaging errors</u>: when finding the average of random errors, the following equation becomes applicable:

$$\sigma_{mean} = \frac{\sigma}{\sqrt{n}}$$

<u>Multiplication of random errors</u>: when finding the product of random errors, the following equation becomes applicable.

$$\sigma_{product} = \sqrt{A^2\sigma_b^2 + B^2\sigma_a^2}$$

Example:
A rectangular parcel of land has the following dimensions, find the area of land.

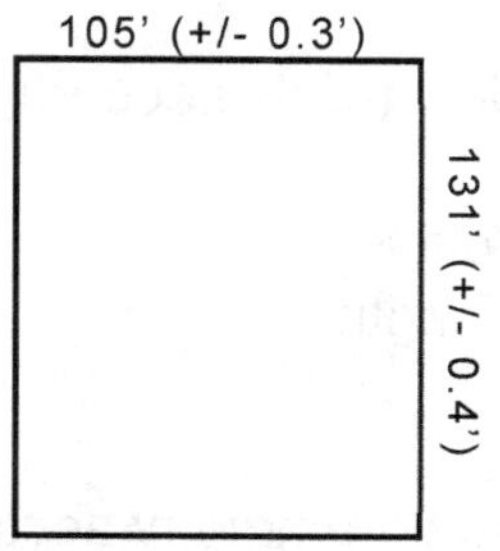

Area is the product of length times width:

$$Area = L \times W$$
$$Area = 105' \times 131'$$
$$Area = 13755 \ square\ feet$$

However, the level of uncertainty is:

$$\sigma_{product} = \sqrt{A^2\sigma_b^2 + B^2\sigma_a^2}$$
$$\sigma_{product} = \sqrt{(length^2)(uncertainty\ of\ width^2) + (width^2)(uncertainty\ of\ length^2)}$$
$$\sigma_{product} = \sqrt{(105^2)(0.4^2) + (131^2)(0.3^2)}$$
$$\sigma_{product} = \sqrt{1764 + 1544.49}$$
$$\sigma_{product} = \sqrt{3308.49}$$
$$\sigma_{product} = 57.52 \ square\ feet$$

The area is 13755 square feet with the uncertainty of +/- 57.52 square feet so, the area is somewhere between 13812.52 square feet and 13697.48 square feet.

FIELD MEASUREMENTS AND CORRECTIONS: ELECTRONIC DISTANCES, STADIA, AND TAPE CORRECTION

Field measurements are subject to nature (e.g., light refraction, temperature, pressure, etc.). This section discusses briefly the information to counteract these acts of nature.

Electronic Distance Measurement (EDM)

V	velocity of light through the atmosphere
c	velocity of light in a vacuum (no medium to hinder the light)
n	index of refraction value (see Snell's Law section in photogrammetry)
λ	wavelength (in meters)
f	modulated frequency in hertz (cycle/second)
D	distance measured
m	integer number of full wavelengths
d	fractional part of the wavelength

As previously discussed in the photogrammetry section in regards to Snell's law, light traveling through a medium (even the atmosphere) will be affected and bent. The equations to resolve these refraction affects are as follows:

$$V = \frac{c}{n}$$

Velocity of light through the atmosphere is equal to the velocity of light in a vacuum divided by the index of refraction value.

$$\lambda = \frac{V}{f}$$

Wavelength (in meters) is equal to the velocity of light through the atmosphere divided by the frequency.

$$D = \left(\frac{m\lambda + d}{2}\right)$$

Distance is equal to the number of wavelengths times wavelength (in meters) plus the fractional part of the wavelength divided by two.

Stadia readings

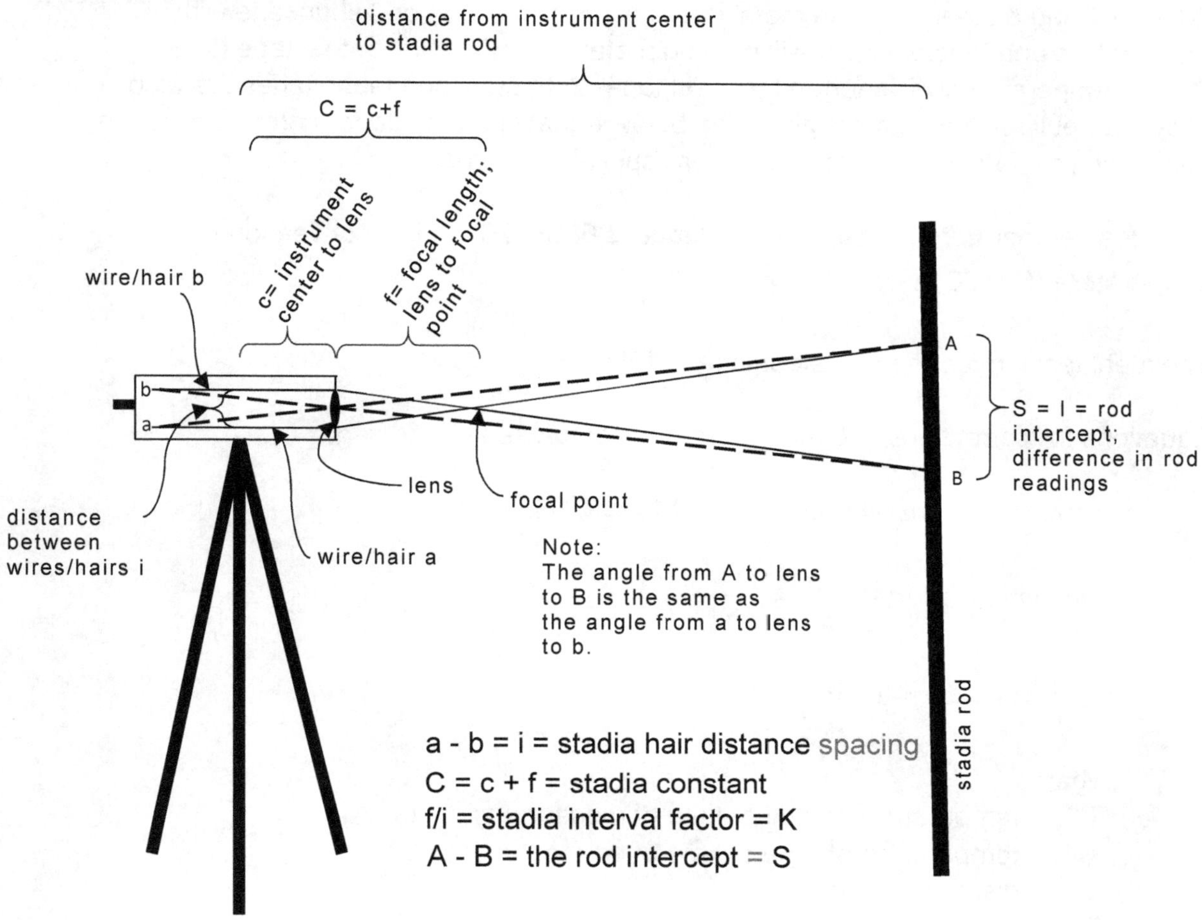

a - b = i = stadia hair distance spacing
C = c + f = stadia constant
f/i = stadia interval factor = K
A - B = the rod intercept = S

$$horizontal\ distance = KS\ cos^2\alpha$$
$$vertical\ distance = KS\ (\sin\alpha)\ (\cos\alpha)$$

Where:

- K   The stadia interval factor and is focal length/stadia hair spacing.
- S   The rod intercept and is the distance between the upper and lower readings on the stadia rod.
- α   The slope angle (measured from the horizontal plane). In the diagram above the slope angle is zero because the instrument is pointing level.

Angle measurement types

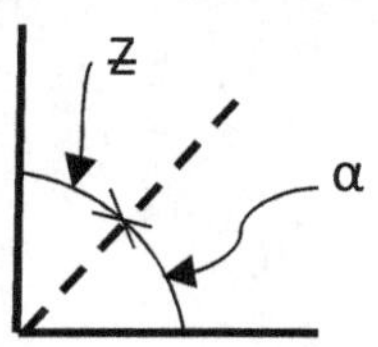

Ƶ The zenith angle, measured from vertical line down.

α The slope angle, measured from horizontal plane up.

Tape measurements and corrections

When utilizing a steel tape, the metal is subject to expansion and shrinkage with temperature and tension. Heat will expand metal and will make a steel tape longer. Colder temperatures will shorten the steel tape. Both steel and fabric tapes are also subject to pull tension and sagging. The below equations provide for corrections when these external forces and temperature are applied to a tape.

The ideal temperature for use of a steel tape is 68°F≈20°C (The conversion for temperature °F to °C is $(°F - 32) \times \frac{5}{9} = °C$).

The ideal tension for use of a steel tape is 12lb.

Equations for corrections in temperature, tension/pressure, and sag:

$$temperature\ correction = C_t = 6.5 \times 10^{-6}(T - T_s)L$$

$$tension\ correction = C_p = \frac{(P - P_s)L}{(AE)}$$

$$sag\ correction = C_s = \frac{(w^2 l^3)}{(24P^2)}$$

Where:

T	temperature of tape at time of measurement (°F)
T_s	temperature of tape at time of correction* (°F)
L	distance/length
P	pull applied at time of measurement ($lbs.$)
P_s	pull applied at time of correction* ($lbs.$)
A	cross-sectional area of tape (inches)
E	elasticity of tape (pounds per square inch-psi)
w	weight of tape (lbs./feet)
l	length of unsupported section of tape (feet)

* time of correction considerations are made at ideal temperatures/tensions or factory default settings.

LEVELING AND A LEVEL LOOP

Leveling is the method of establishing vertical information associated with survey points. The vertical component of a survey point determine the elevations. There are several methods of leveling. Some of the methods to be discussed are as follows: barometric leveling, differential leveling, trigonometric leveling, reciprocal leveling, and precise leveling.

Barometric leveling

The concept of barometric leveling is using the atmospheric air pressure readings at a point's location to determine its elevation. Barometric leveling is not an accurate method of obtaining elevations of points. However, barometric leveling can be utilized for initial or preliminary elevation establishment in a relatively fast manner, especially if the points are large distances apart from one another.

Differential leveling

Differential leveling is the most accurate for of leveling (see the section for total stations, levels, scanners, GNSS and GPS, and UAS). Differential leveling utilizes a horizontally fixed position leveling instrument to obtain backsight (BS) and foresight (FS) readings from a vertical/upright positioned and calibrated measuring pole (known as a stadia rod, leveling rod, grade rod, elevation rod, etc.). Differential leveling requires a leveling instrument that is solely purposed for backsight reading from a previously established point and carrying that vertical information to the newly created foresight reading. The known point elevation is carried up (+) to the level instrument (known as the height of instrument (HI)) with the backsight reading and then brought down (-) to the new point with the foresight reading.

First,

Elevation of BS + reading of BS = HI

Second,

HI – reading of FS = Elevation of FS

Trigonometric leveling

Trigonometric leveling is less accurate than differential leveling, but can still be applied with fairly accurate results, if circumstances guarantee it. Unlike differential leveling the instrument does not have to be horizontally fixed, thus a common total station or theodolite can be used. With trigonometric leveling, a leveling instrument is not utilized. The requirement is the slope distance between the points and the slope angle (angle measured up from horizontal plane) are known to compute a new point's elevation. The concept of trigonometric leveling is triangulation and the use of right triangle mathematics are applied (see the section on triangles).

Example of trigonometric leveling:

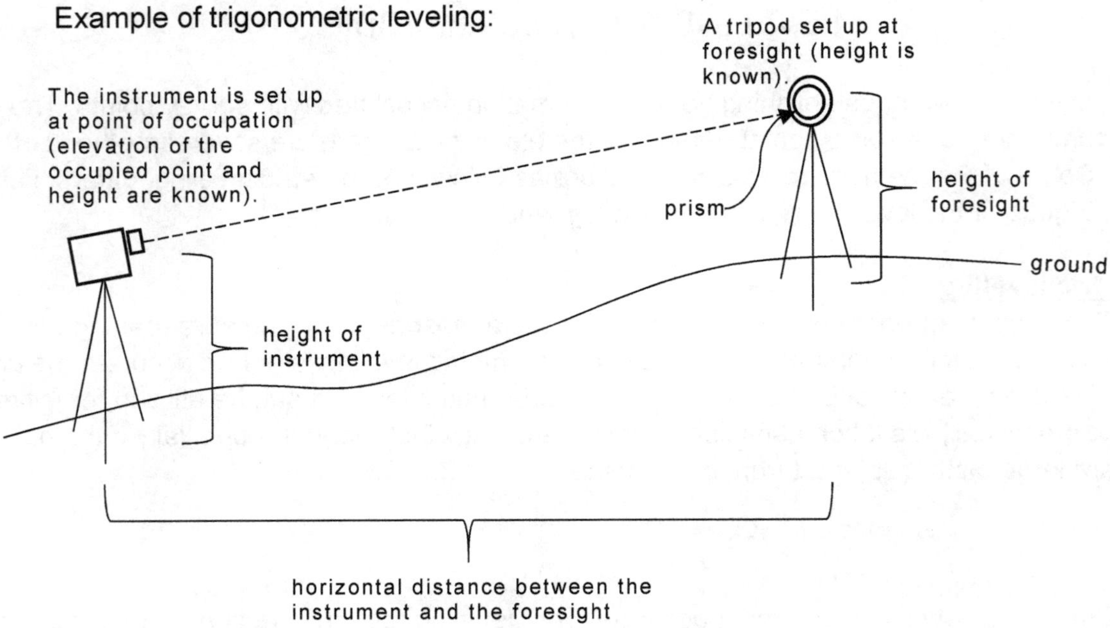

From the information depicted above, a right triangle can be formed and the trigonometry can be utilized to determine the elevation of the foresight.

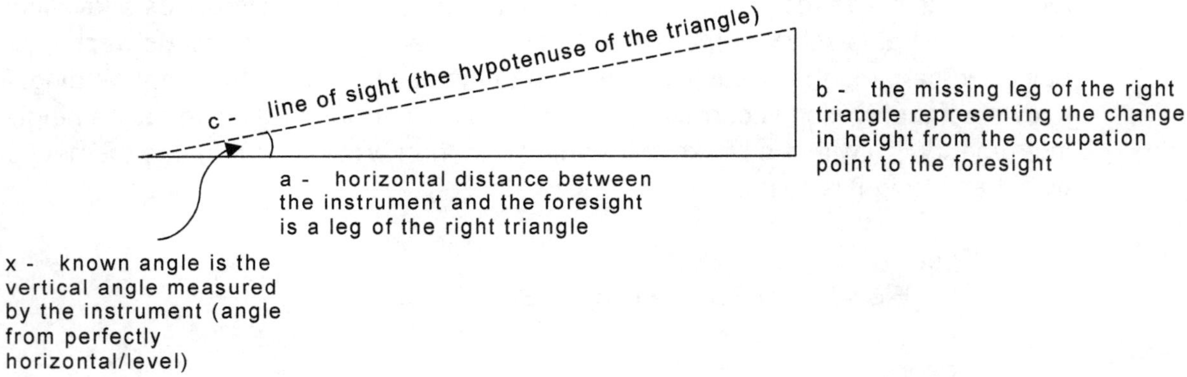

If a=205' (horizontal distance), and c=205.07' (slope distance), then:

$$a^2 + b^2 = c^2$$

$$205^2 + b^2 = 205.07^2$$

$$42025 + b^2 = 42053.70$$

$$b^2 = 28.70$$

$$b = 5.36$$

Example:
If the slope angle (x) is known to be 1.4977° and c=205.07' (slope distance), then the following equation can also be used to find missing value b:

$$\sin x = \frac{b}{c}$$

Type equation here.

$$\sin 1.4977 = \frac{b}{205.07}$$

$$0.0261 = \frac{b}{205.07}$$

$$0.0261 = \frac{b}{205.07}$$

$$(0.0261)(205.07) = b$$

$$5.36 = b$$

In both situations, the change in height from the instrument to the prism on the foresight is 5.36'; the foresight is 5.36' higher in elevation than the occupation point. Note, the height of the tripod at the foresight and the height of the instrument need to be considered and compensated to determine the elevation of the ground at the points.

Reciprocal leveling

Reciprocal leveling is a method of leveling that attempts to compensate for earth curvature and atmospheric errors. Typically, the distance between the points being leveled are far apart. The concept is measurements are obtained close to and far from both points and then averaged.

Example of reciprocal leveling:

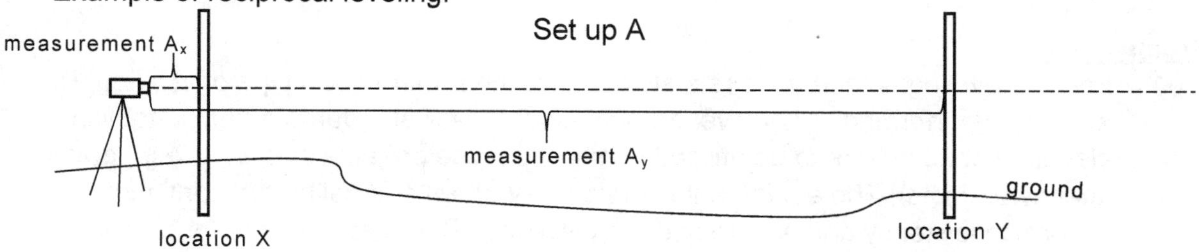

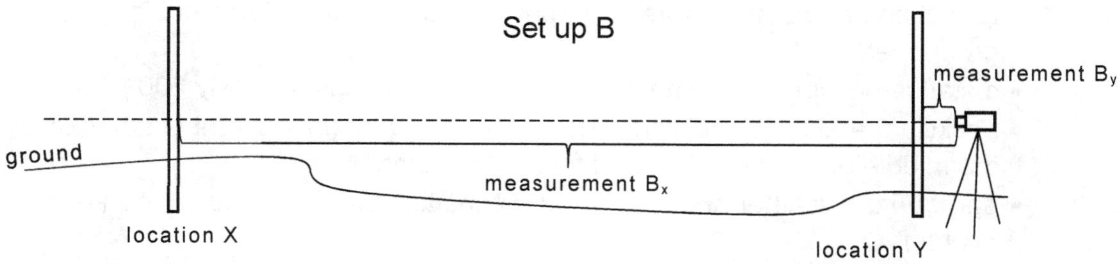

The readings taken to the further distance from each set up are observed at face value; where errors for earth curvature and atmospheric conditions not being addressed. These errors are eliminated if the readings are taken from both sides of the points in the same manner. The errors associated with set up A when reading location Y are cancelled out when the errors associated with set up B when reading location X are observed.

$$\textbf{difference in height between locations X and Y (H)} = \frac{(\boldsymbol{A_y} - \boldsymbol{A_x}) + (\boldsymbol{B_y} - \boldsymbol{B_x})}{\mathbf{2}}$$

Example:
If the height readings while at station A to location X is 1.700' and to location Y is 2.800';
If the height readings while at station B to location X is 1.800' and to location Y is 2.700'.
Then, what is the height difference between locations X and Y?

$$\mathrm{H} = \frac{(A_y - A_x) + (B_y - B_x)}{2}$$

$$\mathrm{H} = \frac{(\mathbf{2.800'} - \mathbf{1.700'}) + (\mathbf{2.700'} - \mathbf{1.800'})}{2}$$

$$\mathrm{H} = \frac{\mathbf{2'}}{2}$$

$$\mathrm{H} = \mathbf{1'}$$

Precise leveling

Precise leveling is more or less a style of leveling rather than a type of leveling. Precise leveling is performed to achieve the highest accuracy standards and to establish elevation benchmarks to be utilized in highly precise projects or efforts (e.g., government agency projects). The equipment used is finely tuned and calibrated continuously to assure consistency and accuracy during leveling. Things to consider for precise leveling may include:

- use of high grade equipment,
- well-adjusted and calibrated equipment,
- reduction of potential errors (e.g., parallax, out of level/plumb equipment, unstable ground, etc.,
- distance between instrument set ups (limited to approximately 300',
- backsight and foresight distances should be equal (instrument set up exactly in the middle of the backsight and foresight stations),
- appropriate weather and temperature conditions (e.g., not during rain or snow events)

Level Loop

A level loop is the process of using a leveling instrument to establish and verify benchmarks/elevations along a closed circuit. The below example field book notes depicts the procedure for a closed level loop. The most important concept is the addition of backsight readings and the subtraction of foresight readings. The instrument setups are positioned either between two turning points or between a turning point and a known benchmark. The readings are generated from the instrument setup location.

Point #/Point Name	Backsight (BS) (+)	Instrument Height (I)	Foresight (FS) (-)	Elevation (ELEV)
BM-99	5.22			100
setup 1		105.22		
TP 1	4.87		3.65	101.57
setup 2		106.44		
TP 2	4.22		5.01	101.43
setup 3		105.65		
TBM 1	5.32		6.11	99.54
setup 4		104.86		
TP 3	4.66		5.1	99.76
setup 5		104.42		
TP 4	3.55		6	98.42
setup 6		101.97		
TP 5	4.1		3	98.97
setup 7		103.07		
BM-99			3.05	100.02

READING SUMMATIONS (∑ BS) AND (∑ FS)	31.94		31.92
MISCLOSURE: (∑ BS) - (∑ FS)	31.94 - 31.92= 0.02		
MISCLOSURE/ SETUP #	0.02/7= 0.002857143		

Definitions

Benchmark (BM)	A permanent position of known elevation.
Turning point (TP)	A temporary position derived from backsighting and/or foresighting.
Backsight (BS)	The back reading from a turning point, carries known elevation to the occupied turning point.
Foresight (FS)	The forward reading from a turning point, transfers the elevation from the turning point to the next position.

Instrument height (HI) Height of the instrument

Level loop adjustment calculation

The example closed level loop on the previous page shows a mis-closure error of +0.02. The distribution of this error throughout the level run will depend on the distances between the turning points and benchmarks. The proportioned cumulative distances between points compared to the overall level run distance will provide the ratio. This ratio can be multiplied by the mis-closure to produce the percent of the mis-closure to be added/subtracted to/from each setup's elevation. Essentially, a single proportioning approach.

It is important to remember that the error or mis-closure value is the opposite of what needs to be added/subtracted to the elevations (e.g., an error of -0.03 means that 0.03 needs to be added, or an error of +0.03 means that 0.03 needs to be subtracted).

Another important aspect of level adjustment is the distance values used to find the ratio cumulate throughout the level run (e.g., if a level run is 1000 feet long and the 10 setups are evenly spaced then the first setup's length is 100 feet, the second setup's length is 200 feet, the third setup's length is 300 feet, and so forth and so on).

Equations

$$Existing\ Elevation + BS = HI - FS = New\ Elevation$$

$$error = initial\ BM\ elevation - calculated\ BM\ elevation$$

or

$$error = \sum(backsight\ readings) - \sum(foresight\ readings)$$

$$correction = (-)\frac{portioned\ cumulative\ distance}{total\ distance} \times error$$

An example of adjusting the previously noted level run is as follows:

Point #/Point Name	Elevation (ELEV)	distance between setups	PERCENTAGE: proportion of total level run distance: cumulative setup length / overall length	CORRECTION: applicable proportion of misclosure: -(0.02) x (proportional distance)	adjusted elevations
BM-99	100				
setup 1		57.32	0.149742679	-0.002994854	
TP 1	101.57				101.5670051
setup 2		50.11	0.280649965	-0.005612999	
TP 2	101.43				101.424387
setup 3		52.31	0.417304527	-0.008346091	
TBM 1	99.54				99.53165391
setup 4		60.17	0.574492542	-0.011489851	
TP 3	99.76				99.74851015
setup 5		59.22	0.729198777	-0.014583976	
TP 4	98.42				98.40541602
setup 6		50.33	0.860680791	-0.017213616	
TP 5	98.97				98.95278638
setup 7		53.33	1	-0.02	
BM-99	100.02				100

total distance of level run is 382.79

Individual distances between each point associated with each setup.

Originally calculated elevation plus percentage of the overall error.

These values are the cumulative distances divided by the overall level run length.
Example: for the 5th setup the distance would be the distance from setups 1 through 5 divided by the total distance.

(57.32 + 50.11 + 52.31 + 60.17 + 59.22)/382.79 = 0.7292

or

279.13 / 382.79 = 0.7292

These values represent the portion of the overall error to be applied to the elevation for correction. These values can be calculated using the formula:
(cumulative distances/overall level run length) x error.

Concept: The percentage of the overall error to be applied to this specific setup.

GEODESY

Geodesy is the science of measuring the shape, volume, and area of Earth and/or large portions of Earth's surface.

Ellipsoid (3D) and ellipse (2D)

Note: Depiction does not have a "z value" for accurate depiction of ellipsoid. An ellipsoid has a "z value" making it a 3D object.

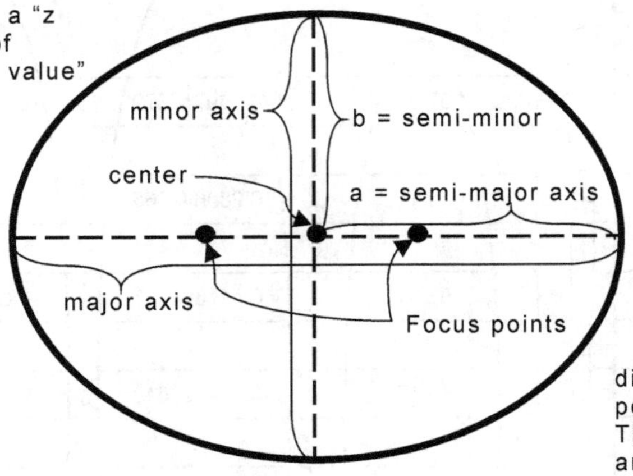

Semi-minor axis = minor axis/2 = b
Semi-major axis = major axis/2 = a

distance from each focus point to center is F. The two foci distances of an ellipse are equal. The equation to find a focus distance is:

$$F = \sqrt{a^2 - b^2}$$

Ellipsoid vs. Geoid

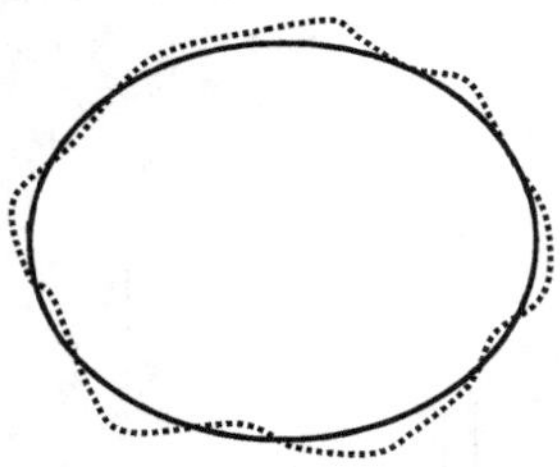

——— Ellipsoid theoretical smooth shape; an ellipse rotated about an axis.

……… Geoid (actual surface) with all the topographic features taken into account.

Flattening and eccentricity of an ellipse

Flattening — Flattening is the measurement of compression on a circle along a diameter.

$$flattening = f = \frac{a-b}{a}$$

Eccentricity — Eccentricity is the ratio of the distance from the center to a focus point and the semi-major axis. If eccentricity is zero, the ellipse is a circle. The greater the eccentricity the more "squished" the ellipse is.

$$eccentricity = e = \sqrt{1 - \frac{b^2}{a^2}}$$

Area of an ellipse

$$ellipse\ area = \sqrt{(area\ of\ major\ axis\ circle)(area\ of\ minor\ axis\ circle)}$$

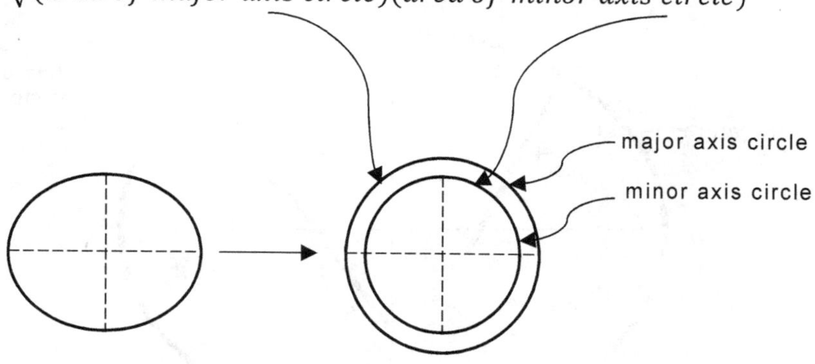

Also: $ellipse\ area = (major\ axis\ radius)(minor\ axis\ radius)(\pi)$

Ellipsoid radius of curvature explanations

The end of the radius (N) always is on the polar axis. The end for the meridian radius (M) is in the quadrant on the other side of the equator. This location has no physical meaning.

Radius in meridian M: Radius does not extend to the y-axis. The radius of this circle is tangent to the ellipsoid at any given latitude and has the same (north-south direction) curvature as the ellipsoid at this point.

$$Radius\ in\ meridian = M = \frac{a(1-e^2)}{(1-e^2\ sin\Phi)^{1/2}}$$

Radius of curvature in prime vertical N: Radius extends to the y-axis.

$$Radius\ in\ vertical = N = \frac{a}{(1-e^2\ sin^2\Phi)^{1/2}}$$

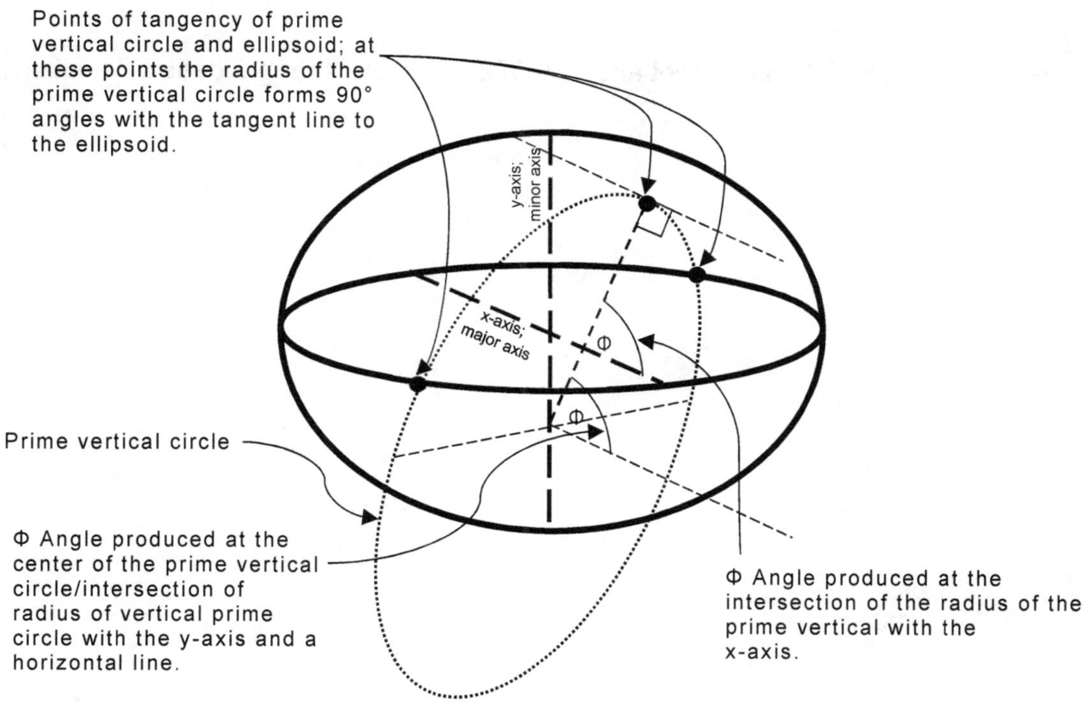

Radius of curvature in meridian and prime vertical comparison

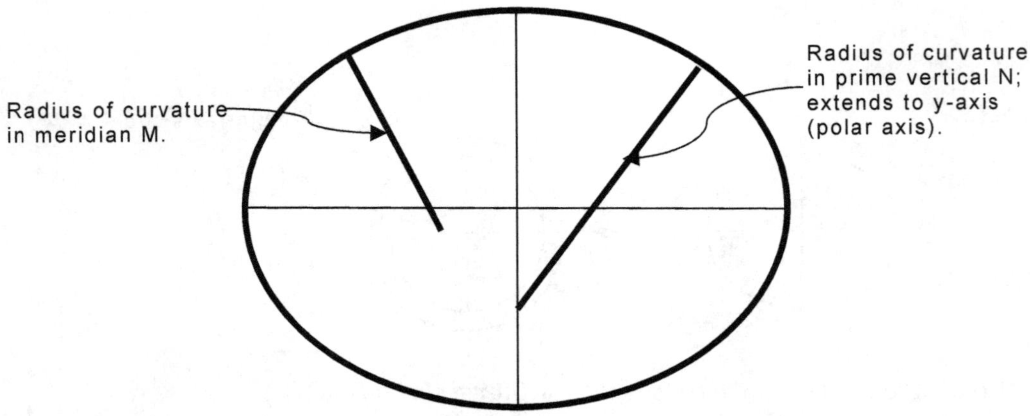

Radius of an ellipsoid at a given latitude

$$radius\ of\ ellipsoid\ at\ a\ given\ latitude = r = \sqrt{\frac{R^2}{1+\left[\frac{1}{(1-f)^2-1}\right]sin^2\Phi}}$$

Where:

R	radius of ellipsoid at Equator
f	flattening
Φ	latitude (angle)

Mean radius of the Earth

Assumed to be 20,906,000 feet and/or 6,372,000 meters for geodetic calculations.

GROUND TO GRID CALCULATIONS

The conversion of ground distances to grid distances and vice versa are an important aspect of land survey calculations. These calculations will mainly be utilized during coordinate system conversions.

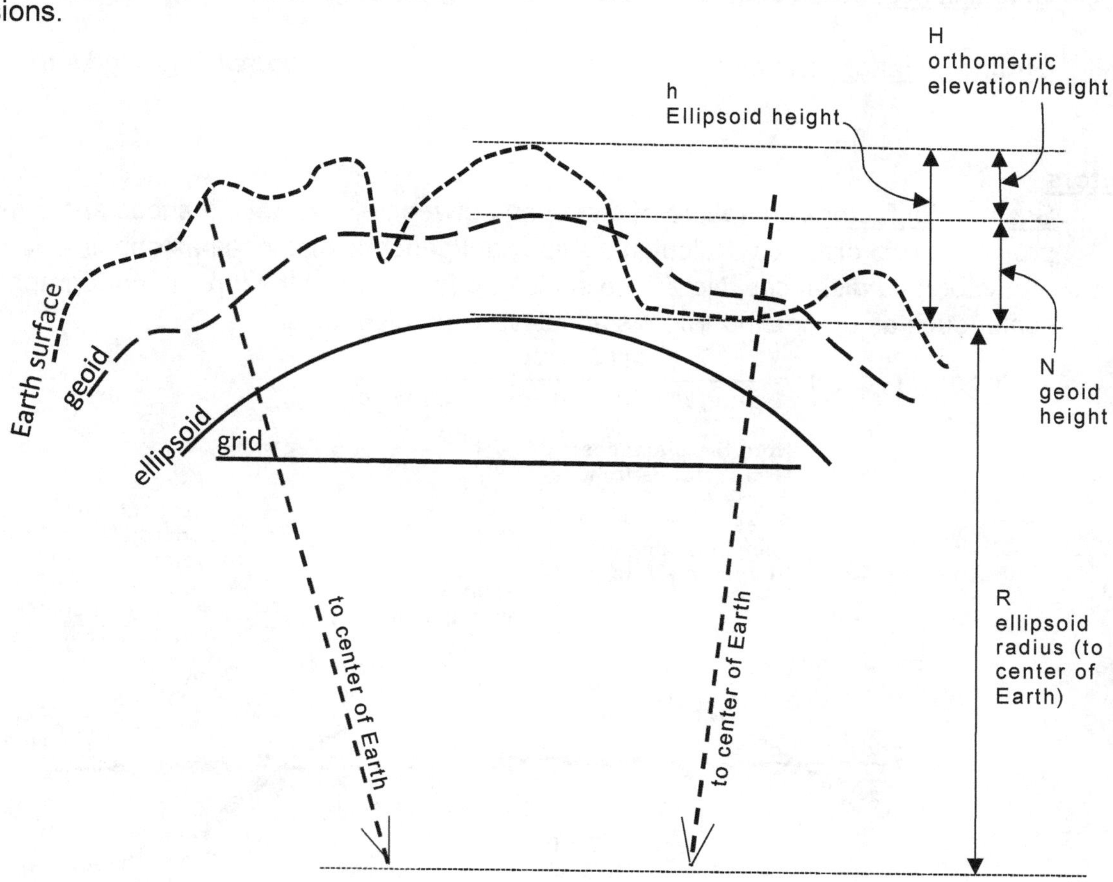

Ellipsoid: A smooth mathematical representation of Earth's surface.
Geoid: A global model of Earth's surface accounting for gravity and rotation.
Grid: A mathematical flat projection of a portion of Earth's surface.
Ground/Topographic: Actual surface of Earth.

Heights and elevations

Ellipsoid radius: Distance from Earth center to ellipsoid surface.

Orthometric elevation/height H: Distance from geoid surface to actual Earth surface.

$$H = h - N$$

Ellipsoid height h: Distance from ellipsoid surface to actual Earth surface.

$$h = H + N$$

Geoid height N: Distance from ellipsoid surface to geoid surface.

$$N = h - H$$

Distances

Ellipsoidal (geodetic) distance: The distance as measured along the surface of the ellipsoid.

Grid distance: The distance as measured along the projected map/grid surface.

Ground distance: The distance as measured along the topographic/ground surface.

Factors

Scale (grid) factor: A measure of distortion between ellipsoidal distances and a map projection/grid distance. Calculates ellipsoidal/geodetic distances as opposed to map projected/grid distances. Note: The scale/grid factor is not the typical "cartographic scale" such as comparing true distances vs. map distances.

$$scale\ (grid)\ factor = \frac{grid\ distance}{geodetic\ (ellipsoidal)\ distance}$$

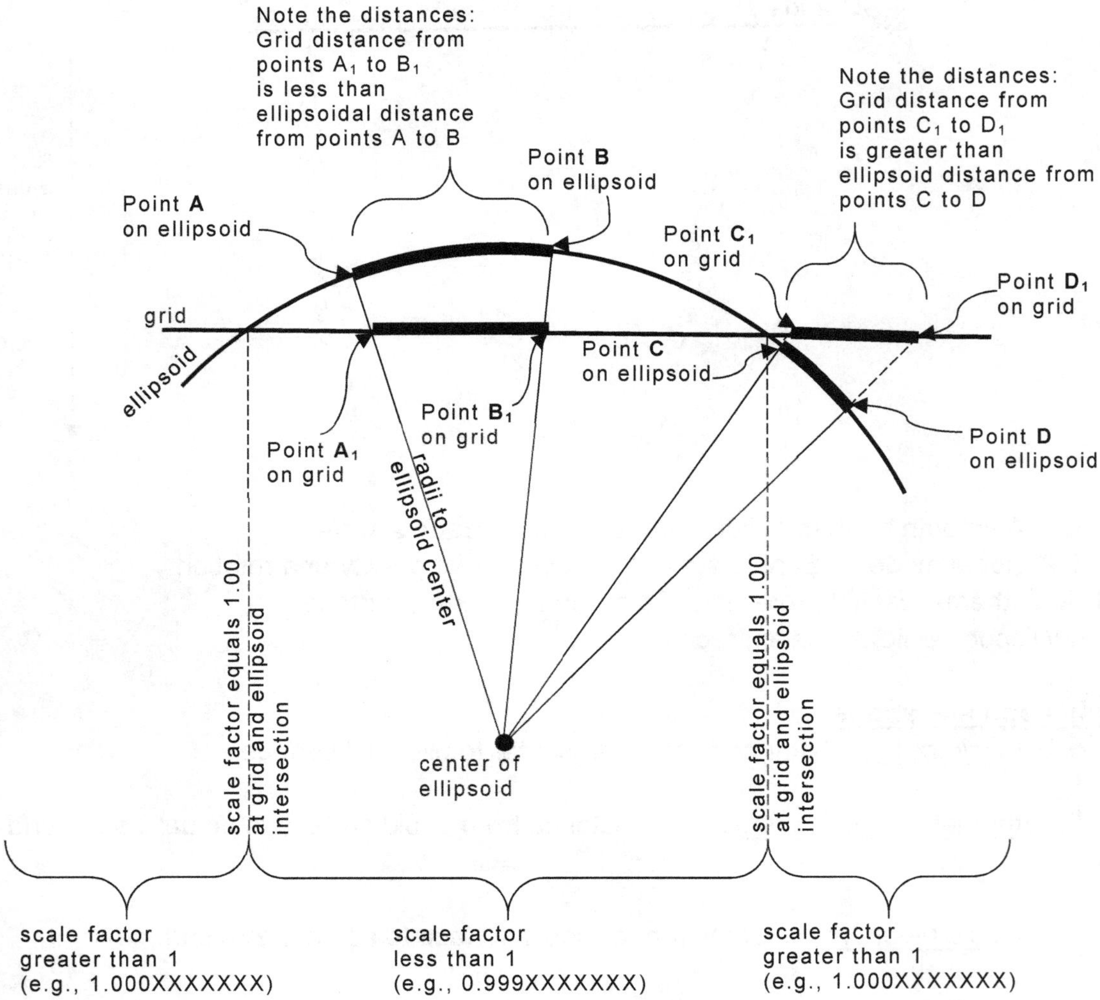

Elevation factor/orthometric height scale: Used in the conversion of distances between the ellipsoid (geodetic) distances and ground distances.

$$elevation\ factor = \frac{R}{(R + H + N)}$$

Where:

R	ellipsoid radius
H	orthometric height
N	geoid height

Combined scale factor: The product of the scale (grid) factor and the elevation factor/orthometric height scale

$$combined\ scale\ factor = scale\ factor\ \times elevation\ factor$$

Ground/Grid conversions:

$$grid\ distance = ground\ distance \times combined\ scale\ factor$$

$$ground\ distance = grid\ distance \div combined\ scale\ factor$$

TRAVERSE ADJUSTMENTS: LEAST SQUARES, COMPASS RULE, AND TRANSIT RULE

Survey traversing is the process of transferring information (horizontal and/or vertical) from one control station to the next throughout a circuit. A circuit can either be a closed circuit, meaning the begin and end stations are the same previously established and known point. The second traverse option is having different known points for the begin and end stations. In either situation, the starting/known information of the control station(s) should be matched with the information obtained from the completed traverse. The occurrence of errors throughout a traverse can result in a mis-closure, meaning the starting/know information of the control point(s) and the ending traverse results do not match. There are ways to adjust and resolve the mis-closure discrepancy, two of the more common methods are compass rule (also known as the Bowditch adjustment) and the transit rule.

Compass rule (also known as the Bowditch Adjustment)

The compass rule solution takes both angular and length adjustment into account, and assumes equal weight between angular and length measurements throughout a traverse.

The compass rule assumes both angles and distances are of equal weight and have the same degree of error.

The main concept is a direct proportionality between latitude/departure and length.

$$\frac{individual\ traverse\ leg\ error\ in\ latitude/departure}{total\ error\ in\ latitude/departure\ of\ traverse} = \frac{length\ of\ individual\ traverse\ leg}{total\ length\ of\ traverse\ perimeter}$$

The process of adjusting a traverse with the compass rule is as follows:

Step 1: Solve for the latitude and departure of the overall mis-closure, solve for $error_{latitude}$ and $error_{departure}$. (see section on double proportioning for latitude and departure calculations).

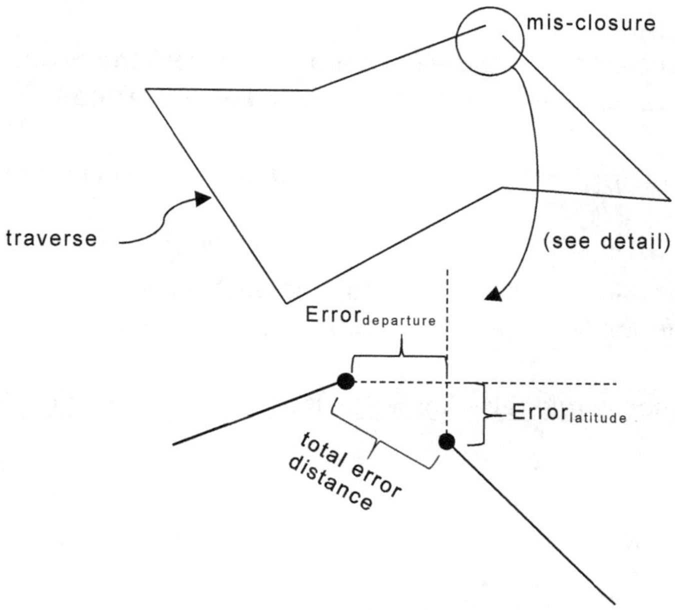

$$overall\ error = \sqrt{(\ error_{Latitude})^2 + (\ in\ error_{Departure})^2}$$

Additionally, precision is calculated by the following:

$$precision = \frac{error_{overall\ closure}}{total\ length\ of\ traverse\ perimeter}$$

Step 2: Solve for the latitude and departure valves for each leg of the traverse.
Very important:

South latitude values are NEGATIVE and north latitude values are POSITIVE
West departure values are NEGATIVE and east departure values are POSITIVE

Next: Solve for the associated correction for each leg of the traverse both in latitude and departure values. The equation for this step is as follows:

$$correction\ in\ latitude_{leg\ 1} = \frac{(overall\ error_{latitude})length_{leg\ 1}}{total\ length\ of\ traverse\ perimeter}$$

The concept here is the correction in latitude for any given leg of a traverse is equal to the product of the latitude of the overall mis-closure of the traverse and the length of the given leg divided by the total length/perimeter of the traverse.

$$correction\ in\ departure_{leg\ 1} = \frac{(overall\ error_{departure})length_{leg\ 1}}{total\ length\ of\ traverse\ perimeter}$$

The concept here is that the correction in departure for any given leg of a traverse is equal to the product of the departure of the overall mis-closure of the traverse and the length of the given leg divided by the total length/perimeter of the traverse.

Step 3: Add /subtract the latitude and departure corrections to each leg's unadjusted latitude and departure values to solve for the adjusted values.

$$adjusted\ latitude_{\ leg\ 1} = (unadjusted\ latitude_{\ leg\ 1}) + (correction\ in\ latitude_{\ leg\ 1})$$

Note:
Correction value will be positive for north and east shifts
Correction value will be negative for south and west shifts

<u>Example of compass rule (also known as the Bowditch adjustment) application</u>

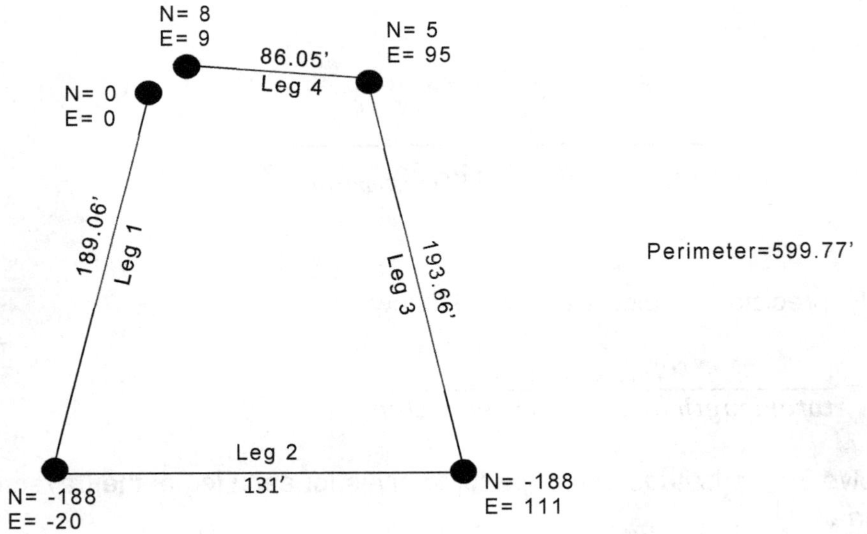

Step 1: Solve for mis-closure latitude and departure values and the overall closure error.

The latitude is the change in northing= -8 (south)
The departure is the change in easting= -9 (west)

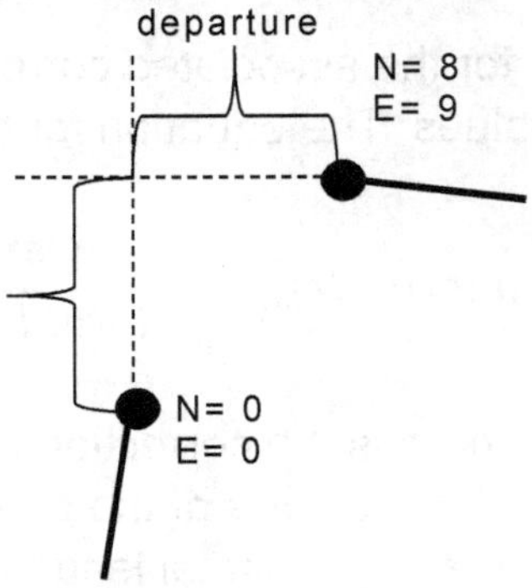

$$overall\ error = \sqrt{(error_{Latitude})^2 + (in\ error_{Departure})^2}$$

$$overall\ error = \sqrt{(8)^2 + (9)^2}$$

$$overall\ error = 12.04$$

AND

$$precision = \frac{error_{overall\ closure}}{total\ length\ of\ traverse\ perimeter}$$

$$precision = \frac{12.04}{599.77} = 0.02007 \approx 1\ in\ 49.815$$

Step 2: Solve for mis-closure latitude and departure values for each leg of the traverse.

Leg 1=	latitude= -188 (south)	departure= -20 (west)
Leg 2=	latitude= 0	departure= 131 (east)
Leg 3=	latitude= 193 (north)	departure= -16 (west)
Leg 4=	latitude= 3 (north)	departure= -86 (west)
Overall latitude/departure	-8 (needs to move south)	-9 (needs to move west)

Note: The overall latitude/departure solved for equals the change in northing and easting from the initial point (N=0, E=0) and the ending point (N=8, E=9). The directions associated with the overall mis-closure need to be preserved FROM ending location TO beginning location. In this case the ending location was north and east of the beginning point, thus it needs to move south (-) and west (-). This is the reason for the -8 latitude and the -9 departure values.

Next: solve for the applicable latitude and departure correction valves for each leg of the traverse.

$$correction\ in\ latitude_{leg\ 1} = \frac{(overall\ error_{latitude})length_{leg\ 1}}{total\ length\ of\ traverse\ perimeter}$$

$$correction\ in\ latitude_{leg\ 1} = \frac{(-8)(189.06)}{599.77} = \mathbf{-2.522}$$

Note: -8 is utilized because the shift from the end point to initial point needs to be in the south direction, which is NEGATIVE.

$$correction\ in\ departure_{leg\ 1} = \frac{(overall\ error_{departure})length_{leg\ 1}}{total\ length\ of\ traverse\ perimeter}$$

$$correction\ in\ departure_{leg\ 1} = \frac{(-9)(189.06)}{599.77} = \mathbf{-2.837}$$

Note: -9 is utilized because the shift from the end point to initial point needs to be in the west direction, which is NEGATIVE.

Using the same method for legs 2, 3, and 4 we get the associated correction values.

Leg 1=	correction in latitude= -2.522	correction in departure= -2.837
Leg 2=	correction in latitude= -1.747	correction in departure= -1.966
Leg 3=	correction in latitude= -2.583	correction in departure= -2.906
Leg 4=	correction in latitude= -1.147	correction in departure= -1.291

Step 3: Adjust each leg with the solved corrections. If the listed corrections in latitude and departure are applied to each leg of the traverse accordingly the following adjusted values result:

$$adjusted\ latitude\ _{leg\ 1} = \left(unadjusted\ latitude\ _{leg\ 1}\right) + (correction\ in\ latitude\ _{leg\ 1})$$
$$adjusted\ departure\ _{leg\ 1} = \left(unadjusted\ departure\ _{leg\ 1}\right) + (correction\ in\ departure\ _{leg\ 1})$$

Leg 1=	latitude= -188	(correction in latitude= -2.522)	adjusted= -190.522
	departure= -20	(correction in departure = -2.837)	adjusted= -22.837
Leg 2=	latitude= 0	(correction in latitude= -1.747)	adjusted= -1.747
	departure= 131	(correction in departure = -1.966)	adjusted= 129.034
Leg 3=	latitude= 193	(correction in latitude= -2.583)	adjusted= 190.417
	departure= -16	(correction in departure = -2.906)	adjusted= -18.906
Leg 4=	latitude=3	(correction in latitude= -1.147)	adjusted= 1.853
	departure= -86	(correction in departure= -1.291)	adjusted= -87.291

Adding up the adjusted latitudes should result in zero

-190.522 + -1.747 + 190.417 + 1.853 = **0.000 correct**

Adding up the adjusted departures should result in zero

-22.837 + 129.034 + -18.906 + -87.291 = **0.000 correct**

Transit rule

The transit rule solution takes both angular and length adjustment into account, but assumes more weight on angular measurements throughout a traverse as opposed to length measurements. Not very applicable when considering modern technology and equipment.

Least squares adjustment

A "best fit solution" is obtained when the sum of the squares of the residuals are minimized. The result reduces the amount of correction to the data set.

Think "BEST FIT" solution.

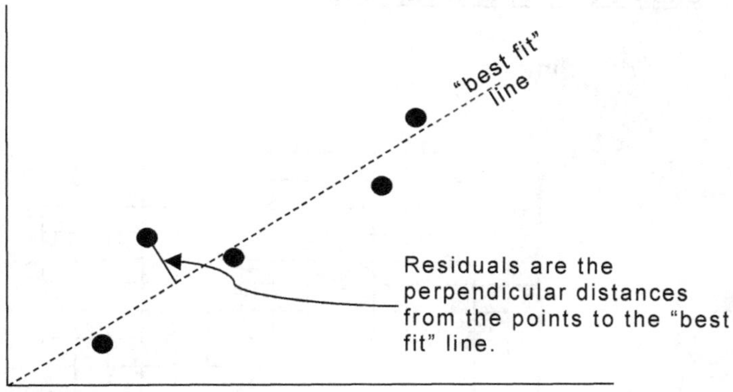

The procedure for least squares adjustments:

In order to determine a solution slope m, y-intercept b, and summations need to occur. The objective is the formulation of a best fit solution while finding the smallest residuals and producing the least amount of adjustment of each individual point. (Refer to the section on intersection of lines for further discussions on slope and y-intercept).

Step 1: For each point, find the value of x^2 and the value of (x)(y).

Step 2: Add up all the values of x, y, x^2, and (x)(y) individually.

Step 3: Calculate the slope m of the line. Remember slope is the change in y values divided by the change in x values ($m=\Delta y/\Delta x$). However, in regards to coordinates, the equation for slope is as follows:

$$m = \frac{(n)\sum(xy) - (\sum x)(\sum y)}{(n)\sum(x^2) - \sum(x)^2}$$

Where: n= number of measurements (coordinates)

Step 4: Calculate the y-intercept b of the line. In regards to coordinates, the equation for slope is as follows:

$$b = \frac{\sum y - (m)\sum(x)}{n}$$

Where: n= number of measurements (coordinates)

Step 5: Insert values in to the slope intercept equation (y=mx+b).

<u>Example of least squares adjustment application</u>

Given the following x,y coordinates:

1,1
2,3
6,5
8,8
9,10

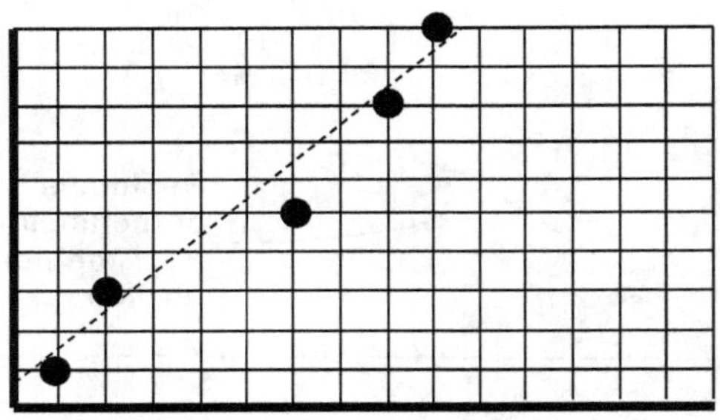

Step 1: Find the values for x^2 and (x)(y) for each point coordinate.

1,1	x^2= 1	(x)(y)= 1
2,3	x^2= 4	(x)(y)= 6
6,5	x^2= 36	(x)(y)= 30
8,8	x^2= 64	(x)(y)= 64
9,10	x^2= 81	(x)(y)= 90

Step 2: Add up the values from each point coordinate.

x, y	x^2	(x)(y)
1, 1	x^2= 1	(x)(y)= 1
2, 3	x^2= 4	(x)(y)= 6
6, 5	x^2= 36	(x)(y)= 30
8, 8	x^2= 64	(x)(y)= 64
9, 10	x^2= 81	(x)(y)= 90
26, 27	186	191

Step 3: Solve for slope m from the coordinates.

$$m = \frac{(n)\sum(xy) - (\sum x)(\sum y)}{(n)\sum(x^2) - \sum(x)^2}$$

$$m = \frac{(5)(191) - (26)(27)}{(5)(186) - (26)^2}$$

$$m = 0.996$$

Step 4: Solve for y-intercept b from the coordinates.

$$b = \frac{\sum y - (m)\sum(x)}{n}$$

$$b = \frac{27 - 0.996(26)}{5}$$

$$b = 0.220$$

Step 5: Now that slope and the y-intercept are known, insert x values into the slope intercept form equation to solve for y values. The "best fit" line is the result of these coordinates.

$y = mx + b$
$y = 0.996x + 0.220$

$y = 0.996(1) + 0.220$
$y = 1.216$
if X=1, then Y=1.216

$y = 0.996(2) + 0.220$
$y = 2.212$
if X=2, then Y=2.212

$y = 0.996(3) + 0.220$
$y = 3.208$
if X=3, then Y=3.208

...so forth and so on. See the graph below depicting the resulting "best fit" line.

Given the following x, y coordinates:
1,1
2,3
6,5
8,8
9,10

Least squares line coordinates found:
0, 0.220 (y-intercept)
1, 1.216
2, 2.212
3, 3.208
10, 10.180

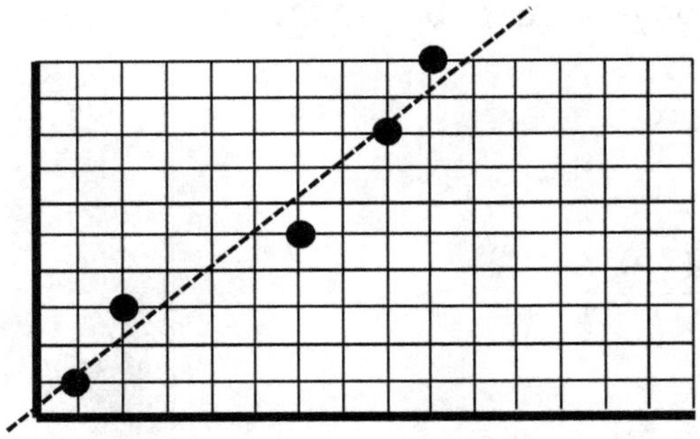

INSTRUMENTATION:
TOTAL STATIONS, LEVELS, SCANNERS, GNSS AND GPS, AND UAS

Land surveying instrumentation has continuously developed as humans attempt to measure land, structures and features, and earth. Surveying instruments have been a part of the construction sites for the Egyptian pyramids until now with modern day homes and high-rise buildings. Throughout the development of these instruments, one common element has not change, their purpose of measuring angles and distances. The accuracy, efficiency, and speed of measuring has changed, but the purpose has remained true.

Some common types of survey instruments are as follows: total stations, levels, scanners, GNSS (Global Navigation Satellite System)/GPS (Global Positioning System), and UAS Unmanned Aerial System).

Total station

The common land survey tool known as the total station is an optical device that measures horizontal angles, vertical angles, and slope distances. Typically, the total station uses an electronic distance measurement (EDM) for making these angular and distance measurements. An EDM requires a glass prism that can reflect the EDM signal back to the total station for measurement. The total station will occupy a previously established control point, will backsight a second previously established control point, and will then perform triangulation calculations to an array of new points. If the bearing and distance between the occupation point and the backsight point is known, then the angle from the backsight point to a new point with the occupation point as the vertex of the angle can be determined. Additionally, the distance from the occupation point to a new point will be measured. From these two elements (the angle and distance), the coordinates of a new point can be determined. This determination is performed with triangulation calculation or coordinate geometry.

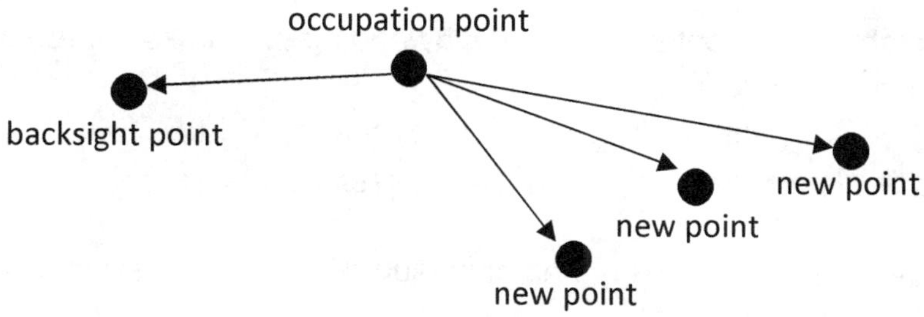

Total stations can be either manual or robotic. Manual total stations require an individual to be physically behind the instrument manually turning and angling the instrument toward another individual carrying a prism and rod. However, a robotic total station is electronically motorized and can follow the prism and rod with remote controls and sensors automatically.

Refer to the section of field measurements and corrections for equations associated with EDM measurements.

Level

Unlike a total station, the level only measures vertical heights and does not measure distances or angular measurements. The level is always set up in a way that allows for digital or optical height measurements through a perfectly horizontal plane extending from the instrument. See the section on stadia reading in the section on field measurements and corrections for a depiction. Also, see the section on level loops for the process on level loop adjustments.

Differential leveling - differential leveling is a very common survey operation for establishing elevations to a series of points throughout a project area. Differential leveling carries the known elevation from a previously established point to other new points by adding backsight readings and subtracting foresight readings.

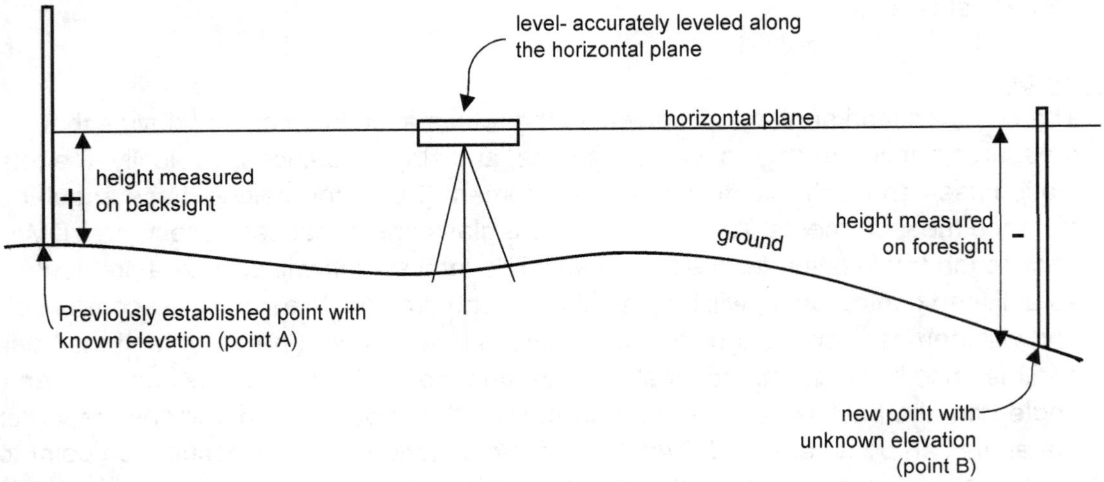

In the depiction above, assume the known elevation of point A is 502 feet. Assume the backsight reading was 3.5 feet and assume the foresight reading of the new point B is 5.5 feet (remember to ADD backsight readings and to SUBTRACT foresight readings).

First, establish the height of the center of the level by adding the backsight reading measurement to the known elevation.

$$502' + 3.5'(\textit{the backsight measurement}) = 505.5'$$
$$(\textit{this is the actual height of the center of the level})$$

Second, carry the elevation to the foresight point by subtracting the foresight reading from the now known height of the level.

$$505.5' - 5.5'(\textit{the foresight measurement}) = 505.5'$$
$$(\textit{this is the ground elevation of the new point } B)$$

Note: The known point elevation is carried up (+) to the level with the backsight reading and then brought down (-) to the new point with the foresight reading.

Peg test - it is always important to calibrate and to test a level prior to use. A common method to test and potentially adjust a level is known as a peg test. The concept is the measurement of elevation difference between two points should remain consistent regardless of the changing location of the level between and around said points.

The process of a peg test is as follows:

Step 1: The distance between the two test points (point A and B) should be approximately 100 feet and the ground between the two test points should be relatively level.

Step 2: Step up the level between the two test points (approximately 50 feet from each point). This is known as position X.

Step 3: Take a height measurement (reading) on point A.

Step 4: Take a height measurement (reading) on point B.

Step 5: Calculate the difference in elevation between the two test points by subtracting the reading on point B from point A.

Step 6: Move the level to a location approximately 10 feet from point A and approximately 110 feet from point B. This is known as position Y.

Step 7: Take second height measurement (reading) on point A.

Step 8: Take second height measurement (reading) on point B.

Step 9: Calculate the difference in elevation between the two test points by subtracting the second reading on point B from the second reading on point A.

Step 10: The resulting calculation from step 6 is compared to the resulting calculation from step 9. The differences in elevations should remain the same.

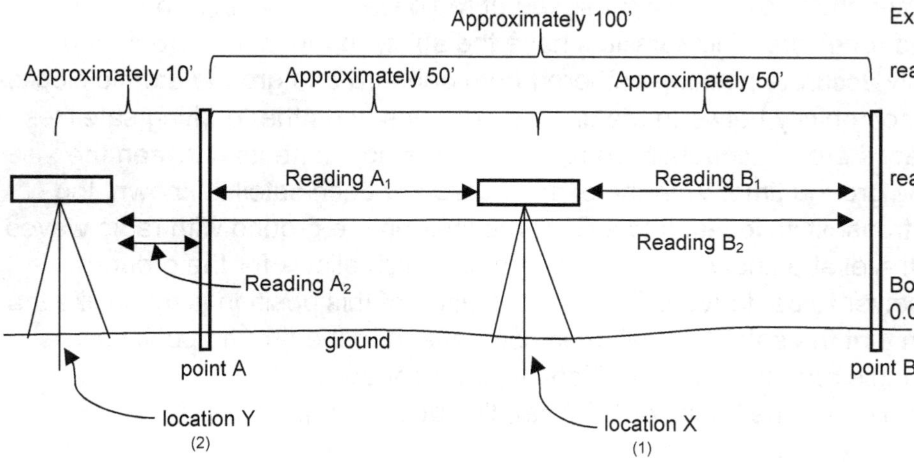

Example:

reading A_1 (1.524) – reading B_1 (1.501)
0.023 is the first result from location X

reading A_2 (1.257) – reading B_2 (1.234)
0.023 is the first result from location Y

Both result in a difference in elevation of 0.023 meaning the level is accurate

Scanners

Survey scanning devices utilize LiDAR (Light Detection and Ranging) technology (also known as 3D laser scanning). The process of scanning focuses on the emission of high speed pulsed light waves from the scanner into the surrounding environment. Light travels at a constant speed, which allows for extremely precise and accurate information retrieval. The light waves scatter out and make contact with features in the surrounding environment. The scattering of the light waves is accomplished with the use of rotating mirrors inside of the scanner. The rotation of the mirrors are on both the horizontal and vertical planes, which allow for 3D data. The light waves then bounce off and return back to the scanner's sensor. The direction and time of each light wave is calculated by the scanner to generate angle and distance measurements between the scanner's sensor and the features. This concept is similar to total station (EDM) calculations, but the difference is the sheer magnitude of data and new points. Scanners have the capability to collect millions of measurements in a very short period of time. The level of detail a scanner can produce of it surrounding environment is extraordinary. The data (e.g., light wave detections of angle and distance calculations collected by the scanner) is known as point cloud data. Further processing or extraction of data features is necessary after a scanning session to make the collected data useable.

GNSS and GPS

GNSS is a more generic term and stands for Global Navigation Satellite System. Earth has several different Global Navigation Satellite Systems. Several countries have their own constellations of satellite circumventing Earth. The United States of America's system is call GPS (Global Positioning System) and contains 31 plus satellites. However, the other Global Navigation Satellite Systems internationally are:

- China has Beidou which contains 35 satellites.
- EU (European Union) has Galileo which contains 24 plus satellites.
- Russia has GLONASS which contains 24 plus satellites.
- India has IRNSS/NavIC which contains 7 satellites.
- Japan has QZSS which contains 4 plus satellites.

The GPS operation is centered on known, highly accurate, timing and location of the orbiting satellites and the known constant speed of radio waves traveling from the satellites to ground receivers. The satellites have the ability to monitor, correct, and transmit their highly accurate position gathered from onboard and ground atomic clocks. This allows for synchronicity between the ground positions and other orbiting satellites. The satellite locations are continuously corrected by time adjustments between the space time and the ground time. With the exact location of each satellite known, the satellite can then transmit its location to a GPS receiver on the ground with radio waves. The radio waves travel at a known and constant rate, which allows for the ground position of the receiver to be derived. The determination of this position is much like that of a total station in that the satellite location in comparison to the ground position is a measurement of angle and distance. Additionally, the concept of
Distance = Rate X Time is the basis for GPS satellite positioning.

There are different radio wave frequencies transmitted from the satellites. The frequencies are L1, L2, L3, L4, and L5. The main two frequencies utilized for civilian survey use are L1 and L2.

Errors are possible with the use of GPS. Some of the errors produced are due to atmospheric effects. Essentially, the more atmosphere the radio wave has to travel through to transmit a signal the more chance of error. If the satellite transmitting the signal is straight overhead, then the less atmosphere the signaling radio wave has to penetrate. If the satellite is closer to the horizon, then the more atmosphere it has to penetrate. Another source of error derives from multipathing. Multipathing is the reflection of the radio wave signals off surrounding features prior to being obtained by the GPS ground receiver. Multipathing can occur, for example, if the GPS ground receiver is located between high buildings, under a bridge, under vegetation, under electric wires, etc. Another error type is ephemeris and clock error. This error is produced by delayed satellite position information at the time of data adjustment. If older or invalid satellite positional information is used to calculate a current ground position then the location will change. The longer the delay potentially creates a greater inaccuracy. Error can be introduce purposely as well with selective availability. This occurs when the military purposely and intentionally introduce error into the satellite position.

The orbiting satellite constellation is ever changing and the position of satellites “overhead” can vary based on location on earth of the receiver and time of day.

Dilution of precision (DOP) is the measure of the strength of a given satellite configuration or geometry. The lower the DOP value the better the precision; the higher the DOP value the worse the precision. Dilution of precision affects the accuracy of data collected by GNSS/GPS. There are three main types of dilution of precision are vertical (VDOP), horizontal (HDOP), and positional (PDOP). The more spread out (closer to the horizon) the available satellites are in the sky the better the DOP. The reasoning is the further separation of satellites results in larger angles of measurement.

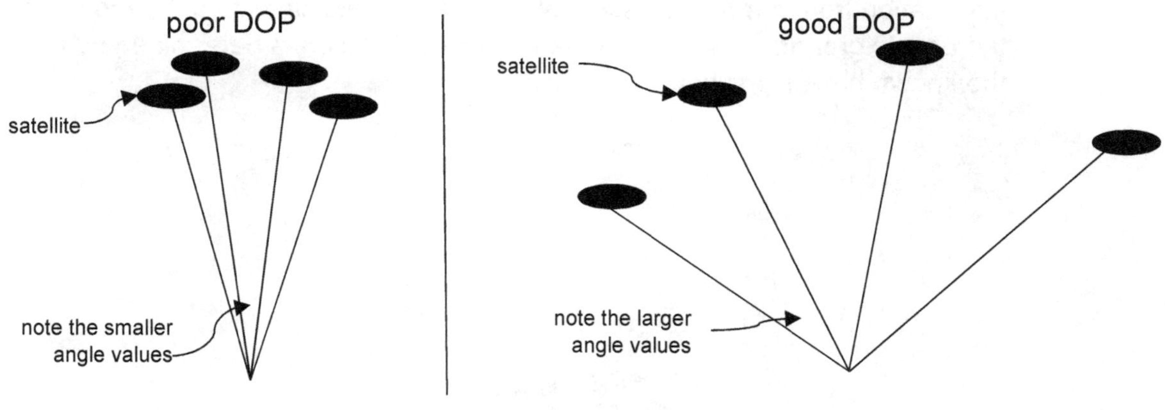

GNSS/GPSS survey types:

Post Processed Kinematic (PPK) - a GPS PPK static survey results in the highest accuracy. Essentially, multiple GPS receivers collect ground positions of multiple locations. These ground position locations produce baselines to each other and allow for redundant measurements between the points. The positions are then adjusted with a least squares adjustment.

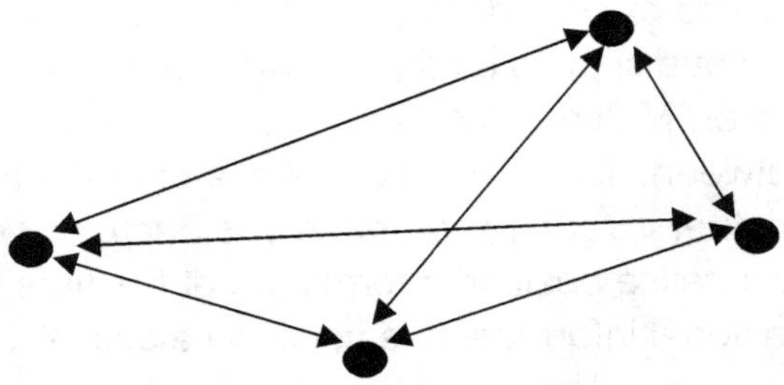

Example:

The line between the points show a baseline.

In this configuration assume all four points are occupied and collect GPS information simultaneously. If this were the case, then any one of the positions can be compared to all three of the other points. The comparison takes the form of a least squares adjustment

Static surveys can take the form of a typical static positioning survey or a fast-static positioning survey. The typical static survey requires 2 hours to 8 or more hours of occupation time of a ground position. The fast-static survey type requires 5 minutes to 20 or more minutes of occupation time of a ground position.

Real-time kinematic survey (RTK) - this type of GPS survey is an "on the fly" type of survey. Typically RTK requires at least two GPS receivers. First, a base receiver that collects the radio wave transmission signals from the satellites (as explained in the above GNSS/GPS section). Secondly, at least one rover receiver. The rover receiver(s) are then given signal and positional information from the base receiver. The rover(s) do not actually get information from the satellites, but rather they get information from the base. Essentially, the base gets satellite data and locks down the base's ground position, and then the rover location is determined by its relation to the known base's position.

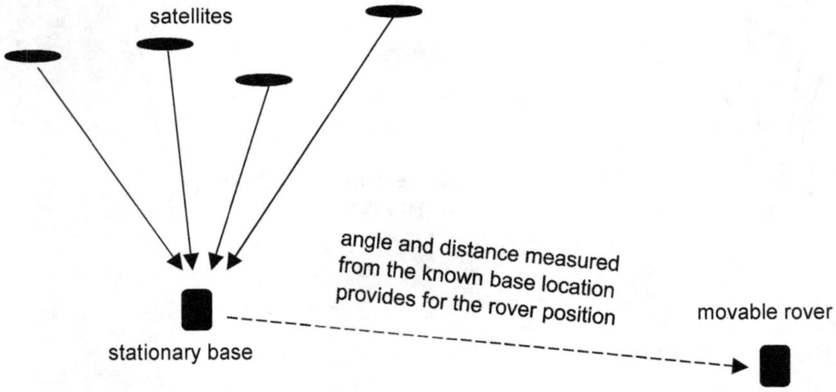

Real-time networks or Virtual reference station (VRS) networks can be established and essentially allow for permanently positioned and maintained base stations to constantly and consistently collect satellite transmission signals and the subsequent positioning of rover receivers.

Online Positioning User Service (OPUS)- is essentially a free service that allows for GPS surveys to be tied into the National Spatial Reference System (NSRS).

One of the purposes of the National Oceanic and Atmospheric Administration (NOAA) is to monitor and observe earth and its system through instrumentation and data collection networks. One of these data collection methods provided by NOAA is the National Geodetic Survey (NGS). The main objective of the NGS is to "define, maintain, and provide access to the National Spatial Reference System (NSRS)" and the NSRS "provides a consistent coordinate system that defines latitude, longitude, height, scale, gravity, and orientation throughout the United States and its territories" (per NGS website- https://geodesy.noaa.gov/).

The Continuously Operating Reference System (CORS) network is a network of real-time kinematic base stations that are continuously feeding GNSS/GPS data and metadata to NGS. The CORS network compiles hundreds of base stations operated by governmental, academic, and private entities. Once the GNSS/GPS data is collected by NGS it is subsequently analyzed and provided to the public free of charge.

With all this said, OPUS utilizes the CORS network to gather and compile data to be provided to NGS and subsequently ties the needed ground position to NOAA's NSRS.

UAS/UAVs

The term UAS stands for Unmanned Aerial System. The term UAV stands for Unmanned Aerial Vehicle. GPS technology can be applied to a UAS. Much like the concept associated with RTK the UAV can serve as a "rover" while being fed information from a base. The UAV's location can be determined "on the fly" and thus information it collects can subsequently be derived. A UAS can collect several types of data such as aerial imagery or LiDAR. As discussed in the above scanner section the LiDAR data is collected in the same manner but from a moving UAV as opposed to a fixed scanner. Aerial imagery can also be collected with the UAV and the same practices as outlined in the photogrammetry section are then applied. In either data collection option ground targets need to be set. The ground targets are set with known horizontal and vertical positions and serve as a reference check when processing the point cloud data or aerial imagery.

TYPES OF SURVEYS

Surveying and the resulting types of surveys can take many forms. Each type of survey has a purpose and requires specific guidelines during their production.

Control survey

Purpose: to establish a series of control ground point positions (horizontal, vertical, or a combination of horizontal and vertical components). These positions will serve as the basis or foundation for subsequent surveys. The use of the points established during a control survey will provide for consistency within a project or amongst several projects and will allow for checking and as a reference for accuracy. It is required that the points established during a control survey are functional (in that they related to one another). The establishment of the points needs to follow a strict and well-documented process. The process can take many forms leveling, traversing, GPS PPK, etc. Depending on the accuracy needed for the control point network, more or less stringent methods can be applied.

Construction survey

Purpose: typically, a construction survey is the act of setting out or collecting survey information in regards to the construction of a feature (e.g., house, dam, high-rise, parking lot, road, pipeline, etc.). During construction, information needs to be staked out. This is known as a construction layout survey. Additionally during construction, the features (e.g., earthwork, structures, etc.) need to be collected for accuracy and documentation purposes. This is called a construction as-built survey. Sometimes, the feature may already exist but it needs to be quantified. An example of this could be a massive wood chip pile produced at a wood processing facility needing to be quantified. This is call a quantity survey.

Topographic survey

Purpose: to establish a depiction of existing features. A topographic survey typically gathers information of a surrounding environment in order to depict its current state. Features being identified and collected during a topographic survey may include ground topography to generate contour lines, building perimeters and heights, pipeline trench edges and depths, trees locations and species, parking lot curbing, road paint striping, etc. Typically, topographic surveys produce the information of an environment's existing state for documentation purposes or for further development and design of said environment.

Land development survey

Purpose: land development surveys can take the form of additional, specific types of surveys such as: subdivision surveys, land use surveys, environmental surveys, and flood plain/wetland surveys. Each sub-type of survey focuses on a specific need for the development of land. Some of these sub-type of surveys are:

Subdivision or platting survey: state, county, and local agencies specific regulations will apply to the subdivision of land. The purpose of a subdivision survey is to layout the lotting of the proposed subdivision and show all required and pertinent needs of the subdivision.

Land use survey: outlines the limits of an areas zoning and use of land limitations.

Environment survey: outlines the environmental features of an area.

Flood plain/wetland survey: outlines the elevations of an area and depicts its proximity to water features to identify flooding hazards or other water related concerns or interests.

FIELD BOOKS AND SURVEY DOCUMENTATION

Field books

Regardless of the type of survey being conducted, it is always recommended that the survey team document and record the actions performed during field operations. Field books are an essential component of the documentation process. Capturing and documenting who, what, when, where, and why is the purpose.

Elements to record and document during a survey in the field book should include:

- Date(s) of survey.
- Name(s) of surveyor in charge.
- Name(s) of other survey personnel.
- Location/address of survey.
- Occupation and backsight point information (point numbers, height of instrument, height of backsight and or rod height, etc.)
- Type of survey being conducted.
- Purpose of survey.
- Diagrams depicting the setting with scale or note of "not to scale" and north arrow.
- Weather conditions (precipitation, wind speed, temperature, humidity, etc.)
- Description of the site regarding potential hazards, slopes, obstacles, vegetation, etc.

Field book information can prove to be invaluable if a surveyor is requested to appear in court. The field book tells the story of who, what, when, where, and why.

Raw data

In addition to a handwritten field book, modern day survey instrumentation collects and records raw data files. Typically, the raw data files are reviewed, manipulated, adjusted, analyzed and so forth and so on. However, it is recommended that a copy of the field collected raw data files be copied and saved in their original state prior to any manipulation. Having a record or version of the original state of the raw data can prove to be very useful for reversion back to the original state, or for proof of original conditions at the time of collection.

MAPPING

Once the field data is collected, analyzed, adjusted, and reviewed, the next step is to incorporate it into a map or some other form of visual depiction. Mapping is the final deliverable in most survey applications.

Main elements of a map or plat:

Title- explains the survey type, general location, and even the purpose of the map.
Example:

Subdivision Plat
Subdivision of Smith Lands
NE/4 of the SW/4 of Section 10, Township 2 South, Range 69 West, of the 6th P.M.
City of Westminster, County of Jefferson, State of Colorado

North or directional arrow- provides the orientation of the map. A north arrow or the compass rose may include a grid north to magnetic north declination conversion angle information (See the section on magnetic declination for further discussion).
Example:

Scale- defines the linear distance between the map depiction and real world distances. (example: 1"=100', 1 inch depicted on the map represents 100' of real world distance.) The larger the scale means less area/space depicted on the map. Likewise, the smaller the scale means more area/space depicted on the map. Scaling can be thought of as a ratio. This can be confusing. A simply a large scale map might depict a neighborhood, while a small scale map might depict a county or state. Large scale maps show smaller areas with greater detail, and small scale maps show larger areas with lesser details.

Example 1:

If the scale is 1"=500', then how many feet are represented by 4.333"

First, line up the inches and feet $\frac{1}{500} = \frac{4.333}{x}$

Second, solve for x with cross multiplication $0.002 = \frac{4.333''}{x}$

$$0.002x = 4.333$$

$$x = 2166.67'$$

Example 2:

Question: what is the largest scale (in multiples of 200') a 3000' x 3000' square building can fit on an 11" x 17" portrait style paper with a 0.5" buffer around the paper needs to be maintained?

Given: 3000' x 3000' square needs to fit on an 11" x 17" piece of paper with 1" buffer, and the scale has to remain a multiple of 200' (i.e., 1"=200', 1"=400', 1"=600', 1"=800', etc.)

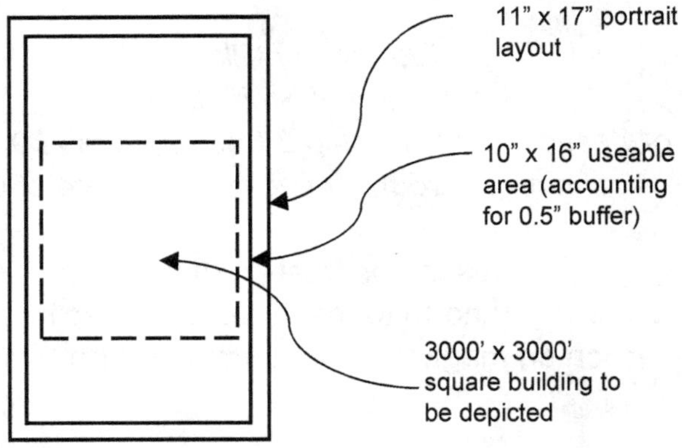

Given the 10" x 16" useable space, the smaller dimension (10") needs to be selected. The 10" dimension will be the dimension that limits the square's depiction, not the 16" dimension.

Try any multiple of 200' scale; 1"=400' scale is selected first in the example for gauging the options.

If 1"=400', then how many inches will 3000' cover?

$$\frac{1}{400} = \frac{x}{3000}$$

$$0.0025 = \frac{x}{3000}$$

$$x = 7.5"$$

If the scale of 1"=400' is chosen, then the 3000' square building would only cover 7.5" of the allotted 10" map space (see below).

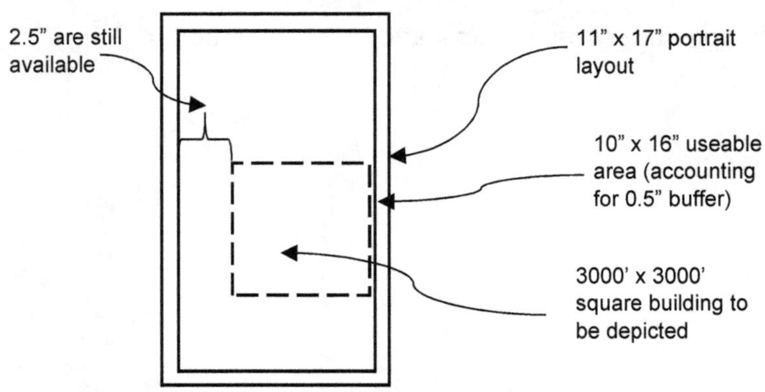

Try the next smaller multiple of 200' scale to see if the 3000' square fits better (remember: smaller scale means larger space used); 1"=200' scale is selected.

If 1"=200', then how many inches will 3000' cover?

$$\frac{1}{200} = \frac{x}{3000}$$

$$0.005 = \frac{x}{3000}$$

$$x = 15''$$

If the scale of 1"=200' is chosen, then the 3000' x 3000' square building would need 15" of the allotted 10" map space (see below).

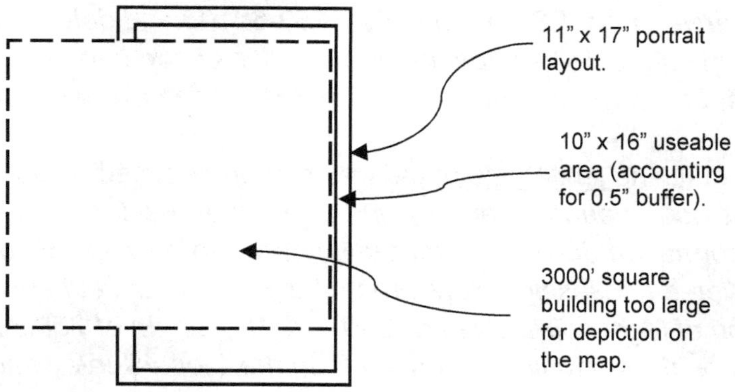

Thus, the scale of 1"=400' would maximize the 3000' square building's depiction on the 11" x 17" map with a 1" buffer.

Legend- provides an explanation of the map's symbols. Points and lines depicted on the map are individually identified in the legend. This allows for fewer labels and explanations throughout the depiction. The legend consolidates all symbol explanations into a single location. Legends can be more or less detailed depending on the map space, clutter, etc.

Example of a Legend:

■ Found Right of Way Monument
○ Found ¾" iron pipe
● Found ½" rebar with red plastic cap (stamped PLS # 23564)
▬ ▬ Fence line
▬▬ Boundary line

Labels- are needed to further identify, depict, and explain features shown on a map that the legend does not or cannot clarify or explain.

Example of a Label:

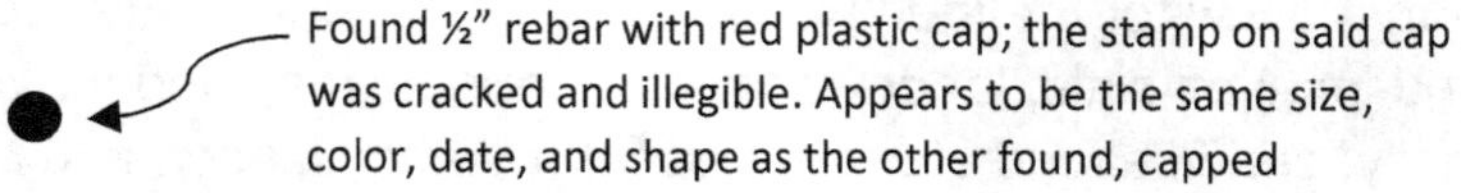

Notes- provide for explanations/clarifications, data source identifications, and limitations to be displayed and made available on the face of a map.

Notes allow for explanations/clarifications of what, why, when, and how the information is depicted on the map, for example: *The northwest no. 5 iron rebar held herein was located within 0.2' from a found, uncalled 1-3/4" galvanized pipe.* Or, *The field data was collected on June 19th, of 2005.* Or, *The verbal testimony of Mr. Cogdill assisted in the recovery of the found 6" x 6" stone monument at the northeast corner depicted herein.*

Notes allow for data source identification of information depicted on the map for example: *The west boundary line is based on the Henry Subdivision Plat as recorded in El Paso County records Plat Book 552, Page 362.* Or, *The flood plain information is derived from FEMA panel 02-8859-62103 and dated 2/28/2013.* Or, *Vesting deed and schedule B-II easement exceptions have been provided by Stewart Title per title commitment no. 12-87963214, effective date 4/12/2017.*

Notes allow for outlining limitations of the information depicted on the map for example: *This is not a boundary survey and is not intended for recording for any purpose, the information depicted is for preliminary use only.* Or, *Evidence of elements of adverse possession were present at the time of field survey, recommendation of legal counsel is advised.* Or, *The length of 258.33' as noted for record distance on the north line is derived from the Siena Survey and distance is illegible on said survey.*

Data sources and citations- provide for the opportunity to identify referenced information as depicted on the map. Much like a works cited, providing the source of all information shown on the map is important, and can allow for further reference. Remember, referenced data and information on the face of a map make that data/information just as valid and useable as all other information shown on the map; a referenced document becomes part of the document.

Vicinity map- allows for a generalized or locative depiction of the overall area being mapped. The vicinity map may show the location of the parcel within the county or even the state (or other large region). Vicinity maps are typically limited to overall features such as major roads, waterbodies, and large regional shapes. Like the map itself, the vicinity map should have a north arrow and scale note.

Details- are essentially the opposite of a vicinity map. A detail is a "blow up" or "zoom in" of a small area of the map that typically would otherwise not be visually possible given the scale of the map. An example of this could be the detail of two found iron pins at a property corner being 3" from each other on a map with an overall 1"=1000' scale. Like the map itself, the vicinity map should have a north arrow and scale note.

Certifications- allow for the identification and statement of authorization for all state, county, and local officials, landowners, land purchases, land surveyors, title researches, etc. Typically, certifications are required for map finalization and authenticity.

Types of a maps or plats:

Plan and profile- this is a type of map that depicts two separate parts or views of a map. The parts or views are tied together in a linear fashion. Typically, one of the views is a top view and the second view is a side view. A plan and profile view can be used to show a pipeline, road, sewer line, etc. location and then its depth below ground.

Example of a plan and profile:

Depicts the linear location of a pipeline and the depth changes that occur with the pipeline as it is bored underneath a river are both depicted.

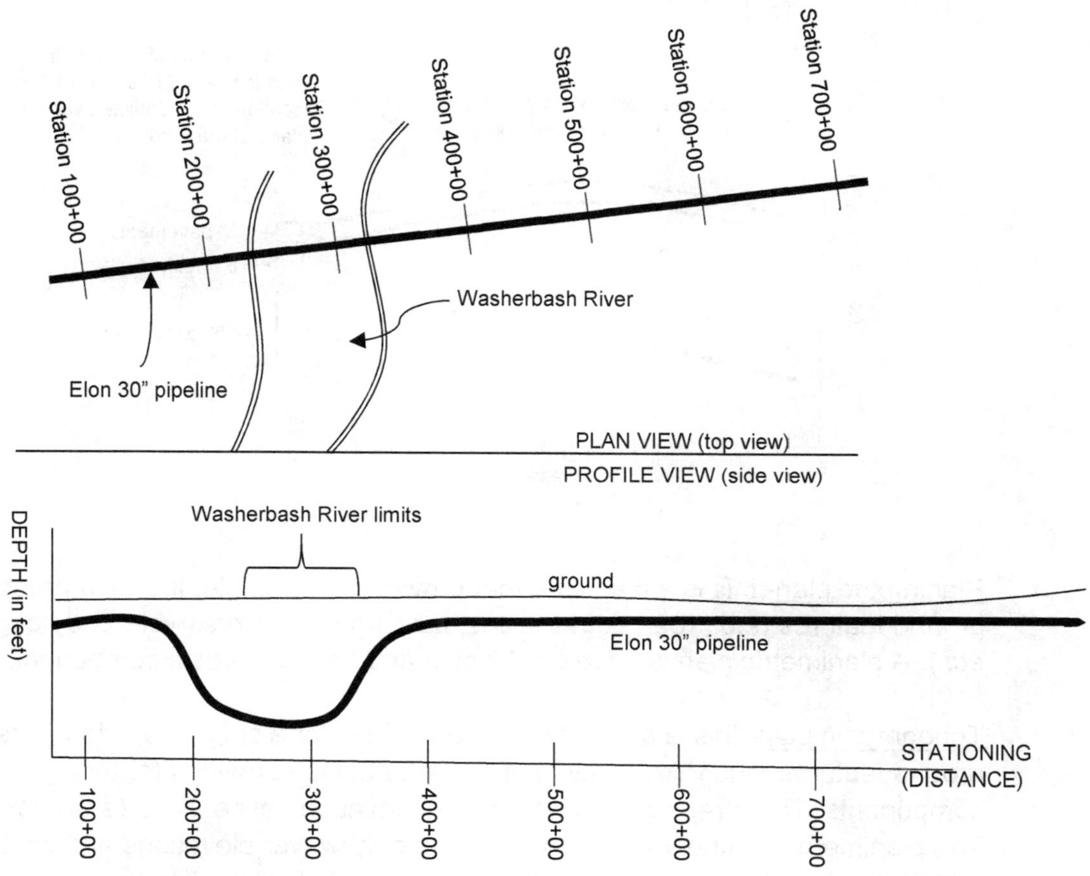

Cross section plan- this is a plan type that provides for a vertical plane cut view of a feature. A cross section plan needs to be tied linearly to the feature it is sectioning, much like a plan and profile. A cross section plan set of a roadway is the best way to show the bed underlay materials and their associated depths.

Example of a cross section plan:

> The vertical plane cut through a road at each 50' interval along the road is depicted; it shows the pavement, gravel, and compacted dirt associated with the road's construction.

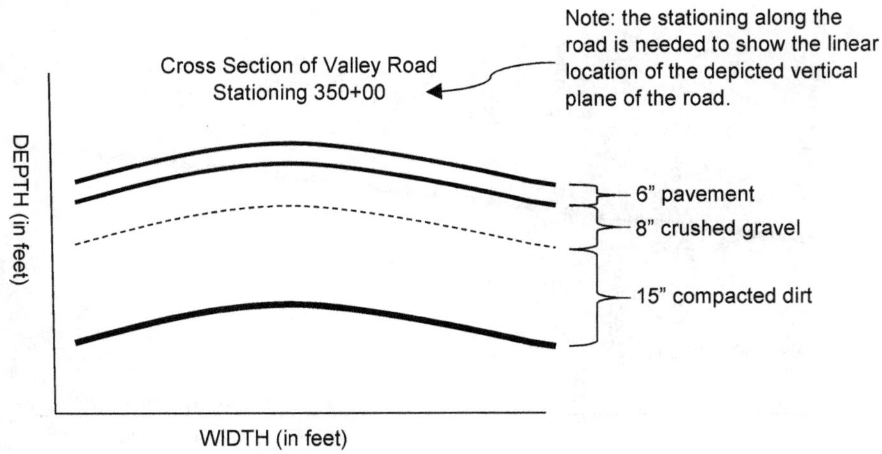

Planimetric plan- this is a plan type that provides for a single, top view perspective of ground features (e.g., road edges, trees, building perimeters, waterbody edges, curbing, etc.). A planimetric plan is a typical 2D plan and has no vertical component.

Topographic plan- this is a plan type that provides for a single, top view perspective of ground features much like a planimetric plan, but also provides for topographical components. Topographic components are elevation delineations (a 3D component). The planimetric components are still available, however elevations and contours are added. (see the section on contours and TINs for further explanation).

Survey plans- there are various survey plans and each type has a specific use and purpose.

Subdivision plan- used for depiction of a simultaneous conveyance of several lot divisions from a larger parent tract. Subdivision plans are typically governed by a public entity (i.e., county or local entities) and require specific information to be depicted. Each public entity has specific requirements that dictate lot placement, lot sizes, lot counts, lot shape, road frontage, road placement and widths, open spaces, common spaces, etc.

Boundary plan- used for the depiction of boundary lines and the analysis results of a professional land surveyor. A boundary survey can and should depict the following elements: parcel boundary limits and notes, adjoining parcel information and notes, bearings and distances (both measured and record dimensions), landowner names, found or missing monumentation, title information, field survey dates, evidence of occupation lines, evidence of unwritten rights (e.g., adverse possession and prescriptive rights), evidence of easements and other encumbrances, recorded easements and other encumbrances, general location, etc.

Mortgage inspection survey- used to identify a parcel being conveyed. The identification should include property lines, public entity setback limits, and improvement features (houses, sheds, pools, porches, fencing, etc.) on the parcel in relation to the property lines and setback limits. A mortgage inspection survey is typically not a boundary survey and is used to discover major issues, but does not define boundary line.

ALTA/ACSM survey- is a standardized survey plan administered by the American Land Title Association (ALTA) and the American Congress of Surveying and Mapping (ACSM). An ALTA/ACSM survey combines many types of survey plans including boundary surveys, mortgage inspection surveys, and topographic and planimetric surveys. As part of the standardization, ALTA/ACSM provides a very specific guideline of items that need to be included on an ALTA survey and explains each item's purpose and method of incorporation into the survey plan. The list of items is identified as ALTA/ACSM Table A. An ALTA survey is a survey that guarantees the information outlined in a provided title commitment. A title company researches and produces a title commitment outlining owner information, vesting deed (current deed) information, chain of title/ownership of the property, current property description, and easements and other encumbrances associated with the property. Once the title commitment is completed, it is presented to a surveyor and the surveyor is in charge of verifying, clarifying, and depicting the information as presented in the title commitment. The number of items selected from the Table A is up to the insurer(s) of the property. An ALTA survey may contain 1 item or all 19 depending on the needs of the insurer(s).

The minimum standard detail requirements for ALTA/NSPS Land Title Surveys (Table A items) most recently released effective February 23, 2021 source: https://www.nsps.us.com/page/2021ALTA) are as follows

NOTE: Whether any of the nineteen (19) items of Table A are to be selected, and the exact wording of and fee for any selected item, may be negotiated between the surveyor and client. Any additional items negotiated between the surveyor and client must be identified as 20(a), 20(b), etc. Any additional items negotiated between the surveyor and client, and any negotiated changes to the wording of a Table A item, must be explained pursuant to Section 6.D.ii.(g). Notwithstanding Table A Items 5 and 11, if an engineering design survey is desired as part of an ALTA/NSPS Land Title Survey, such services should be negotiated under Table A, Item 20.

If checked, the following optional items are to be included in the ALTA/NSPS LAND TITLE SURVEY, except as otherwise qualified (see note above):

1. _____ Monuments placed (or a reference monument or witness to the corner) at all major corners of the boundary of the surveyed property, unless already marked or referenced by existing monuments or witnesses in close proximity to the corner.

2. _____ Address(es) of the surveyed property if disclosed in documents provided to or obtained by the surveyor, or observed while conducting the fieldwork.

3. _____ Flood zone classification (with proper annotation based on federal Flood Insurance Rate Maps or the state or local equivalent) depicted by scaled map location and graphic plotting only.

4. _____ Gross land area (and other areas if specified by the client).

5. _____ Vertical relief with the source of information (e.g., ground survey, aerial map), contour interval, datum, with originating benchmark, when appropriate.

6. _____ (a) If the current zoning classification, setback requirements, the height and floor space area restrictions, and parking requirements specific to the surveyed property are set forth in a zoning report or letter provided to the surveyor by the client or the client's designated representative, list the above items on the plat or map and identify the date and source of the report or letter.

_____ (b) If the zoning setback requirements specific to the surveyed property are set forth in a zoning report or letter provided to the surveyor by the client or the client's designated representative, and if those requirements do not require an interpretation by the surveyor, graphically depict those requirements on the plat or map and identify the date and source of the report or letter.

7. _____ *(a) Exterior dimensions of all buildings at ground level.*
(b) Square footage of:
_____ *(1) exterior footprint of all buildings at ground level.*
_____ *(2) other areas as specified by the client.*
_____ *(c) Measured height of all buildings above grade at a location specified by the client. If no location is specified, the point of measurement shall be identified.*
8. _____ *Substantial features observed in the process of conducting the fieldwork (in addition to the improvements and features required pursuant to Section 5 above) (e.g., parking lots, billboards, signs, swimming pools, landscaped areas, substantial areas of refuse).*
9. _____ *Number and type (e.g., disabled, motorcycle, regular, and other marked specialized types) of clearly identifiable parking spaces on surface parking areas, lots, and in parking structures. Striping of clearly identifiable parking spaces on surface parking areas and lots.*
10. _____ *As designated by the client, a determination of the relationship and location of certain division or party walls with respect to adjoining properties.*
11. *Evidence of underground utilities existing on or serving the surveyed property (in addition to the observed evidence of utilities required pursuant to Section 5.E.iv.) as determined by:*
_____ *(a) plans and/or reports provided by client (with reference as to the sources of information)*
_____ *(b) markings coordinated by the surveyor pursuant to a private utility locate request.*
Note to the client, insurer, and lender – With regard to Table A, item 11, information from the sources checked above will be combined with observed evidence of utilities pursuant to Section 5.E.iv. to develop a view of the underground utilities. However, lacking excavation, the exact location of underground features cannot be accurately, completely, and reliably depicted. In addition, in some jurisdictions, 811 or other similar utility locate requests from surveyors may be ignored or result in an incomplete response, in which case the surveyor shall note on the plat or map how this affected the surveyor's assessment of the location of the utilities. Where additional or more detailed information is required, the client is advised that excavation may be necessary.
12. _____ *As specified by the client, Governmental Agency survey-related requirements (e.g., HUD surveys, surveys for leases on Bureau of Land Management managed lands). The relevant survey requirements are to be provided by the client or client's designated representative.*
13. _____ *Names of adjoining owners according to current tax records. If more than one owner, identify the first owner's name listed in the tax records followed by "et al."*
14. _____ *As specified by the client, distance to the nearest intersecting street.*
15. _____ *Rectified orthophotography, photogrammetric mapping, remote sensing, airborne/mobile laser scanning and other similar products, tools or technologies as the basis for showing the location of certain features (excluding boundaries) where ground measurements are not otherwise necessary to locate those features to an appropriate and acceptable accuracy relative to a nearby boundary. The surveyor must (a) discuss the ramifications of such methodologies (e.g., the potential precision and completeness of the data gathered thereby) with the insurer, lender, and client prior to the performance of the survey, and (b) place a note on the face of the survey explaining the source, date, precision, and other relevant qualifications of any such data.*
16. _____ *Evidence of recent earth moving work, building construction, or building additions observed in the process of conducting the fieldwork.*
17. _____ *Proposed changes in street right of way lines, if such information is made available to the surveyor by the controlling jurisdiction. Evidence of recent street or sidewalk construction or repairs observed in the process of conducting the fieldwork.*
18. _____ *Pursuant to Sections 5 and 6 (and applicable selected Table A items, excluding Table A item 1), include as part of the survey any plottable offsite (i.e., appurtenant) easements disclosed in documents provided to or obtained by the surveyor.*
19. _____ *Professional liability insurance policy obtained by the surveyor in the minimum amount of $___________ to be in effect throughout the contract term. Certificate of insurance to be furnished upon request, but this item shall not be addressed on the face of the plat or map.*
20. _____ ____

Applications for mapping:

Computer Aided Drafting (CAD) - there are several different software applications and programs that are utilized for mapping. Some of the main survey industry standard brands are AutoCAD and MicroStation. CAD allows for the digital creation of all the components required in map production. CAD focuses on the use and manipulation of points and lines (arcs). Survey data is collected in the field and imported in the CAD application. From the imported point data information, lines connecting the points are drafted. Further analysis and review of the data is made accessible by measuring, manipulation, calculation, and adjusting the data. Finally, a map can be generated once the data (points and lines) are finalized and approved. CAD information can be displayed in any type of map projection or coordinate system. This concept of the generation, manipulation, and management of data associated with physical features is known as Building Information Modeling (BIM). CAD applications assist in the performance of BIM.

Geographic Information System (GIS)- is a platform that allows for the capture, analysis, and visualization of spatial and geographic features. Once data is gathered (via typical survey methods as outlined in the instrumentation section), it can be input into GIS for further analysis. Mapping of various features throughout a designated space allow for the analysis of patterns, rates, inventory, conditions, and many other elements of the features being analyzed. The GIS applications produce query lists of the features which allows for comparisons, relations, sorting, and searching for further analysis. The power of GIS is the ability to compare and analyze the data query lists of any given feature(s) with the spatial and geographic location of said feature(s). The applications of GIS is nearly infinite, some examples may include:

1. Mapping out fire hydrant conditions, colors, locations, usage, proximity to other features, etc. throughout a region.
2. Ambulance routes throughout a city analyzed by road speed rates, stop signs, obstacles, etc.
3. As-built pipeline information.
4. Grocery store proximity to neighborhoods.
5. Criminal sex offender mapping.

The GIS spatial and geographic information can be displayed in any type of map projection or coordinate system as defined by the user. The GIS metadata is the information about the collected data. This is the who, what, when, where, how, and why associated with the data (e.g., *The fire hydrant top bolt condition collected along Owen Street was conducted by Blair Madison on 12/5/2020, by use of VRS connected Trimble R12 RTK.*)

Spreadsheets- are a very important part of surveying efforts. Point data is typically organized in ASCII file formats having the file extensions of .TXT, .CSV, and others. An ASCII file is a text file that includes alphanumeric text (letters and numbers) and specifically placed punctuation. The punctuation is placed in a specific manner to dictate a specific result when reading or processing the data (e.g., a comma may represent a separation of information, a hyphen may represent combination of information, etc.). These functions are specific to different file types and programs reading the data. A text file associated with data can easily be sorted, organized, and searched. A field of the data is typically a column in the data set (e.g., all the elevations, all the point numbers, etc.). Examples of user operations are: search for the point with the largest northing (furthest north point), sort out all specifically described points (find all the light poles), organize and sort the points by point number to see the order of field collection, search out all points above or below a certain elevation, and so forth and so on. All of the sorting, organizing, and searching operations are conducted within a spreadsheet. Spreadsheets present data in organized rows (horizontal line of information) and columns/fields (vertical line of information). A row contains all the information associated with an individual item of the data set (e.g., a specific point in the data set). A column or field contains information from the data set items associated with a specific category of the data (e.g., all point numbers, all elevations, etc.). Rows are associated with individual items and columns are associated with categories of all items.

Column or field of data;
a category of information.

Row of data;
item information.

Point Number	Northing	Easting	Elevation	Description
1000	1695024.227	3150853.04	5247.809	140inE yel cap_subject
1001	1695036.195	3150847.088	5248.089	167.5inNE yel cap_subject
1002	1695273.561	3150920.263	5245.925	Yel cap
1003	1695323.404	3150839.878	5244.436	orange cap
1004	1695373.311	3150839.955	5244.384	Red cap
1005	1695423.326	3150839.807	5243.373	Red cap
1006	1695422.992	3150706.205	5240.026	Reb in conc no cap
1007	1695197.974	3150707.063	5243.308	Orange cap
1008	1694905.131	3150870.573	5249.059	Rng pt hole
1009	1695100.726	3150816.255	5247.532	Blg porch cor
1010	1695110.151	3150752.377	5246.88	Blg cor
1011	1695156.326	3150712.747	5245.191	Conc
1012	1695156.338	3150736.319	5245.173	Conc
1013	1695128.582	3150724.641	5245.205	Conc
1014	1695128.575	3150712.8	5245.341	Conc
1015	1695118.617	3150724.652	5245.815	Conc
1016	1695118.408	3150709.397	5244.599	Conc at fence
1017	1695115.282	3150709.577	5244.707	conc at fence

DIGITAL ELEVATION MODELING AND CONTOURS

A Digital Elevation Model (DEM) is a 3D graphic representation of elevation data. A DEM is a generalized term for Digital Terrain Modeling (DTM) and/or Digital Surface Modeling (DSM).

<u>Types of DEMs:</u>

Digital Terrain Modeling (DTM) represents the bare ground elevation data, but additionally can include other natural and man-made features. A DTM can be interpolated to produce a bare ground DEM (the features can be extracted from a DTM to produce a bare ground DEM). However, a bare ground DEM cannot produce a DTM.

Digital Surface Model (DSM) represents a blanketed canopy elevation model that covers both bare ground and natural/man-made features. True bare ground elevations are only observed where features are not present and not observed where obstructed.

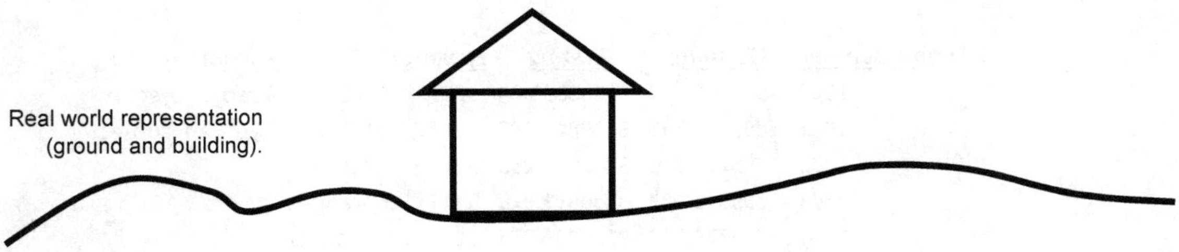

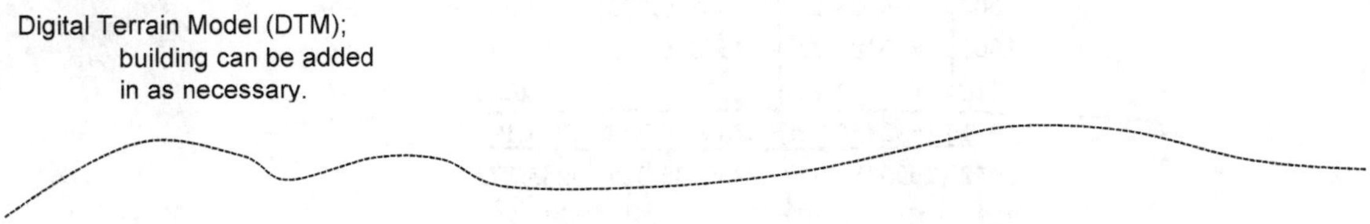

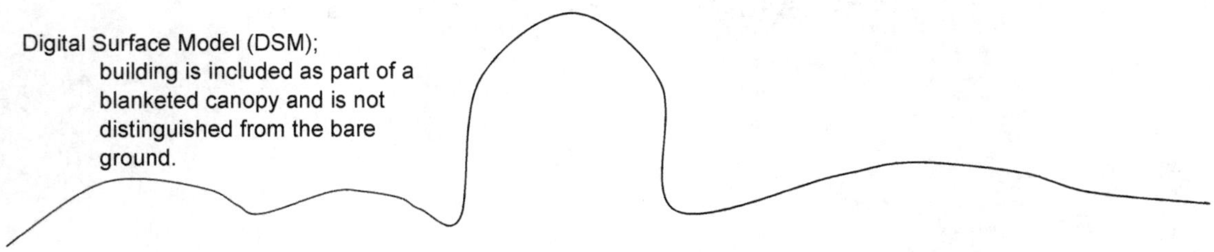

Triangulated Irregular Network (TIN)

The process of generating a DEM can be performed using rasterized images or the use of Triangulated Irregular Networks (TINs). A TIN uses point and line data to produce vectors (lines) between points throughout the terrain. These vectors create a triangular meshwork across the terrain that is interpolated. The triangles formed by the CAD system are generated with the use of algorithms to be as equilateral (similar triangle side lengths) and as equiangular (similar triangle internal angle measurements) as possible. The vertices of the triangles are the point data and the triangle sides are connecting lines between said points. Having the triangles as equilateral and as equiangular as possible essentially means the connecting lines between the point data are minimized. Delaunay triangulation is an approach creating a triangular meshwork having the resulting triangles as equilateral and equiangular as possible. Delaunay triangulation is essential the solution of having a circle drawn through any three data points while not including any other data points.

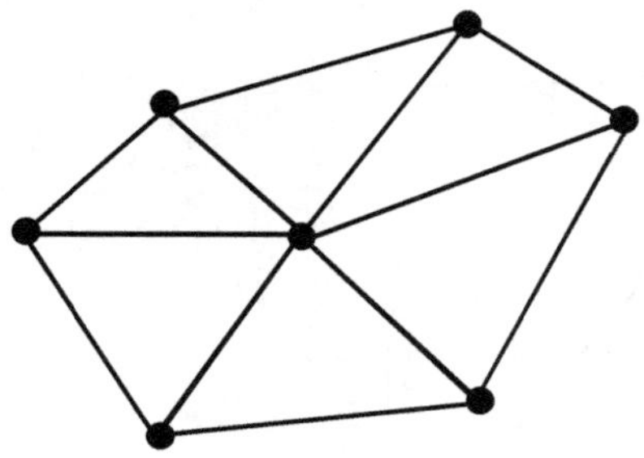

TIN representation showing the points as the vertices of the triangles and having the triangles as equilateral and equiangular as possible.

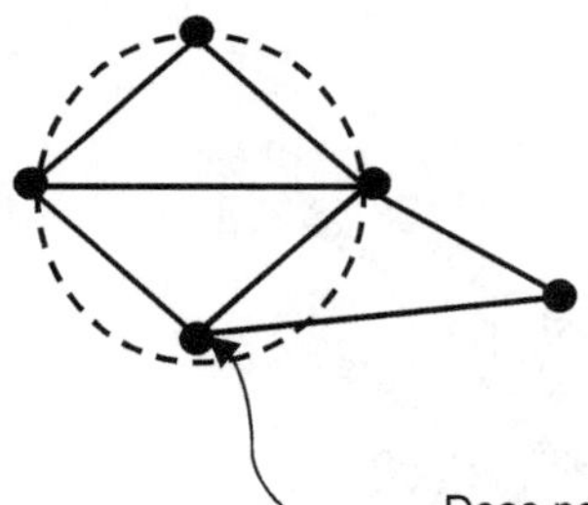

Does not meet the Delaunay triangle approach requirement. The point noted lies within the circle connecting the three points above it.

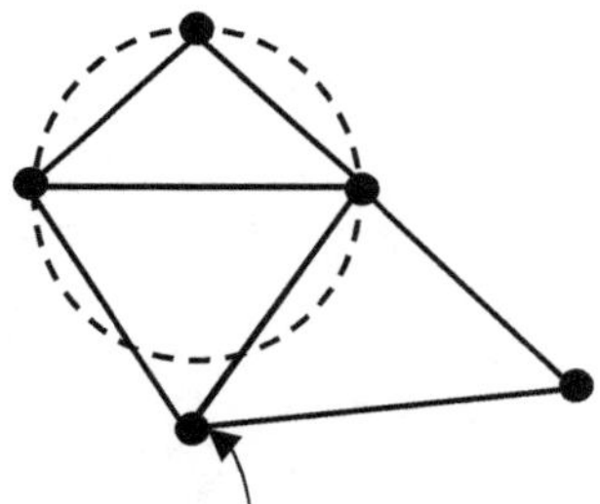

Does meet the Delaunay triangle approach requirement. The point noted lies outside the circle connecting the three points above it.

Interpolation and contouring
Interpolation is the process to deriving contours from the point and vector triangle meshwork. Given the point elevations and the horizontal distances between the point data along the vectors contours can be interpolated.

The process of contour interpolation is as follows:

Step 1: Define a contour interval. The smaller the contour interval the more detail can be obtained, as opposed to a larger contour interval providing less detail.
For this example the contour interval will be set at 100'.

Step 2: Create the tick marks along the available vectors (lines between the points). The tick marks are to be at the previously established contour interval and are to be evenly spaced between the point data along the given vector.

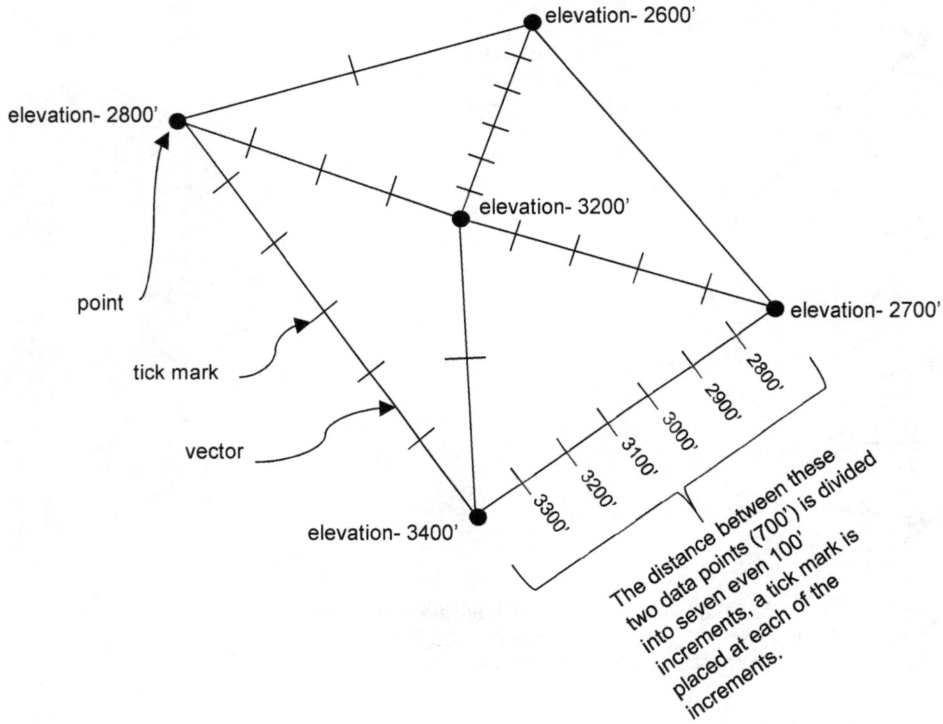

Step 3: Create a smooth contour line connecting the previously created tick marks for each contour interval.

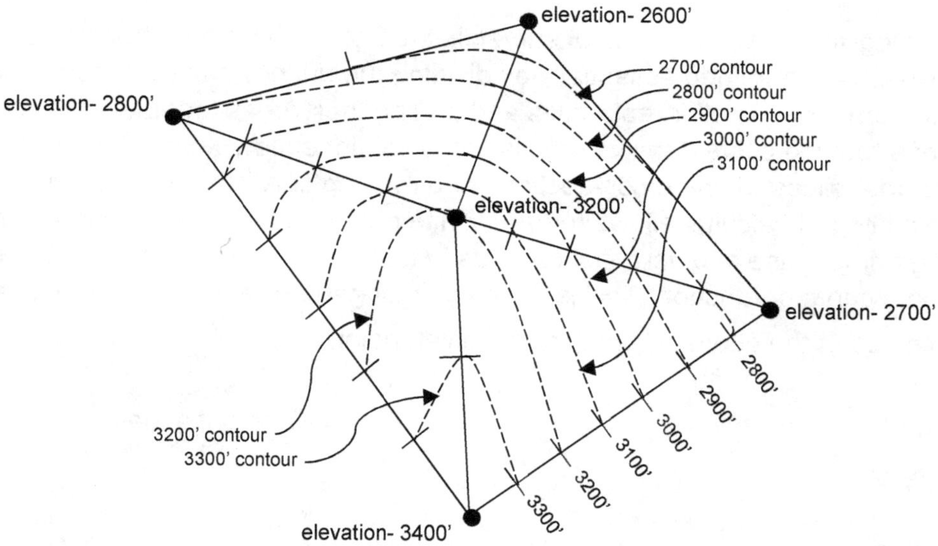

Things to note during contour interpolation should include the following:

- Contours should always point upstream in valleys or flowing water features (i.e., streams and rivers). They should point in the direction of increasing elevations.
- Contours should always point downridge along ridges. They should point in the direction of decreasing elevations.

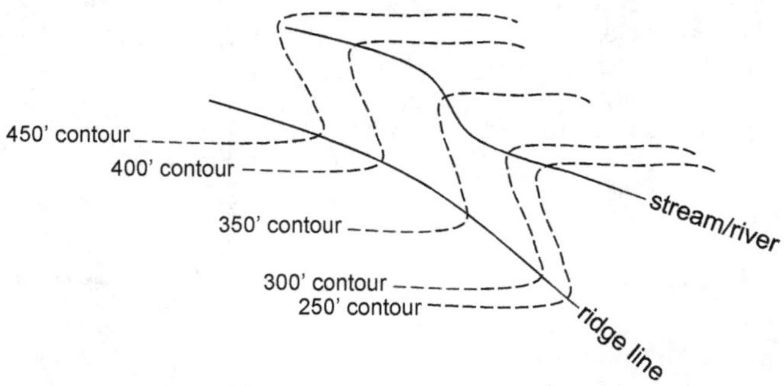

- Adjacent contours should be sequential or equivalent in their interval designation.
- Contours should never divide or split into two separate contours.
- Contours should never cross or loop one another (unless there is an overhanging cliff type feature).
- Contours should never spiral.
- Contours should never stop in the middle of a map.

CALCULUS:
GRAPHED FUNCTIONS

The primary use of integral calculus in terms of survey is the finding of an area beneath a curve or function. The answer is determined essentially by dividing the unknown area beneath the curve into an infinite number of smaller rectangles and adding their areas together. The same applies to volume of a rotating curve if the volume is cross-sectioned into an infinite number of "slices". The smaller the "slices" of the cross-sections of a rotating curve the more accurate the solution. This is what integral calculus allows for; an infinite number of very small parts being added together. The primary use of differential calculus in terms of survey is the finding of slope at a specific point on a curve or function. This is done by finding the rate of change or slope ($\frac{\Delta y}{\Delta x}$) of a very small piece of the curve or function at a given point.

Integration   Differentiation

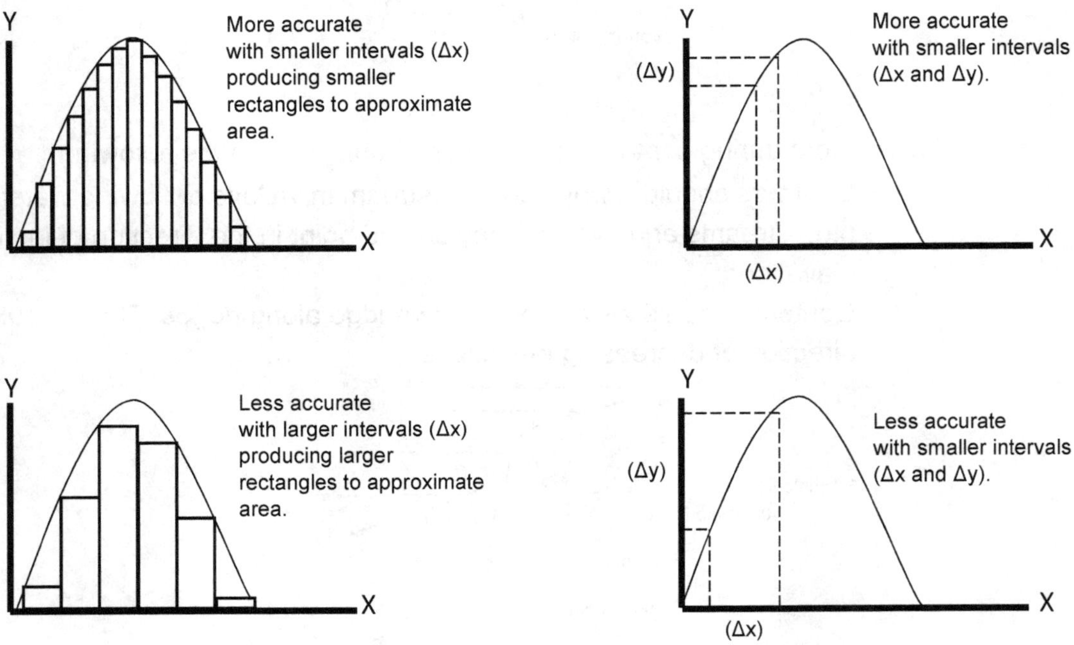

Understanding basic functions

A basic understanding of different types of functions is necessary for understanding of calculus integration and differentiation.

In an equation for a graphed function there are x values and y values corresponding to the line(s) or parabola(s) plotted on the Cartesian graph. The function is written where $f(x)=y$ (e.g., $f(x)=x^2$ or $y=x^2$). From a function, any x value can be determined from any y value, likewise any y value can be determined from any x value.

Example: $f(x) = x^2$ (in other terms $y = x^2$)

$f(x) = x^2$ (if x values are $-2, -3, -4, -5, 2, 3, 4,$ and 5)

If x = 2; $f(x) = 2^2 = 4$; thus $(y = 4)$

If x = 3; $f(x) = 3^2 = 9$; thus $(y = 9)$

If x = 4; $f(x) = 4^2 = 16$; thus $(y = 16)$

If x = 5; $f(x) = 5^2 = 25$; thus $(y = 25)$

If x = -2; $f(x) = -2^2 = 4$; thus $(y = 4)$

If x = -3; $f(x) = -3^2 = 9$; thus $(y = 9)$

If x = -4; $f(x) = -4^2 = 16$; thus $(y = 16)$

If x = -5; $f(x) = -5^2 = 25$; thus $(y = 25)$

If $f(x) = x^2$, then

X	f(x) (also known as Y)
2	4
3	9
4	16
5	25
-2	4
-3	9
-4	16
-5	25

If graphed out the shape is as follows for the function:

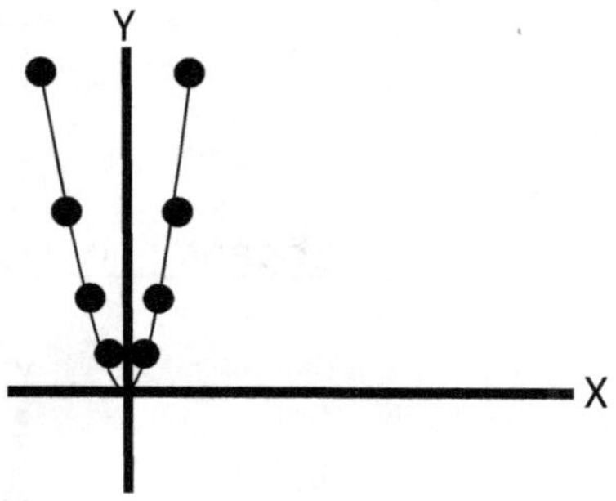

The eight most common graphed function types are as follows (remember the term f(x) equals y):

<u>Linear:</u> $f(x) = mx + b$

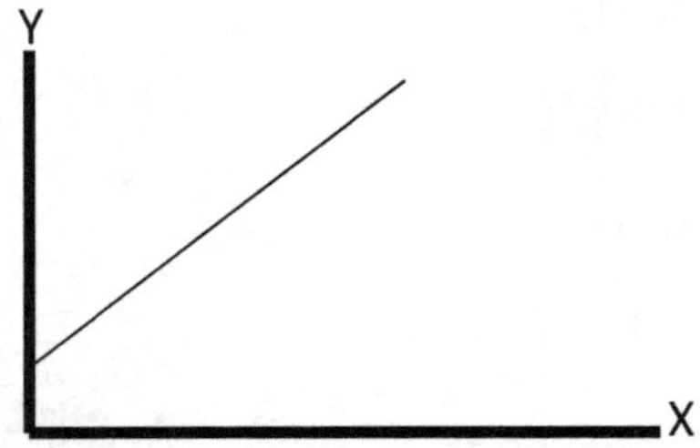

<u>Absolute value:</u> $f(x) = |x|$

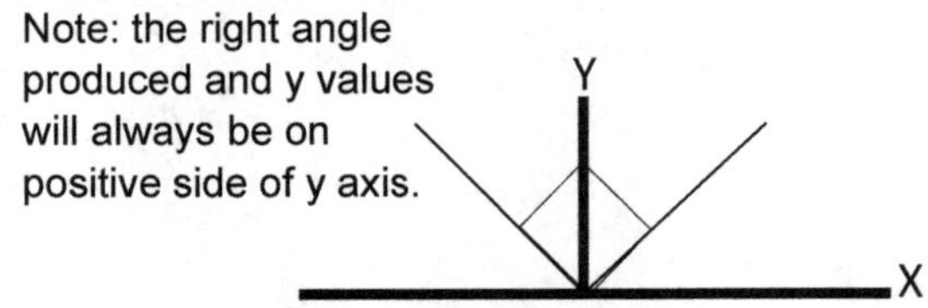

<u>Squared:</u> $f(x) = x^2$

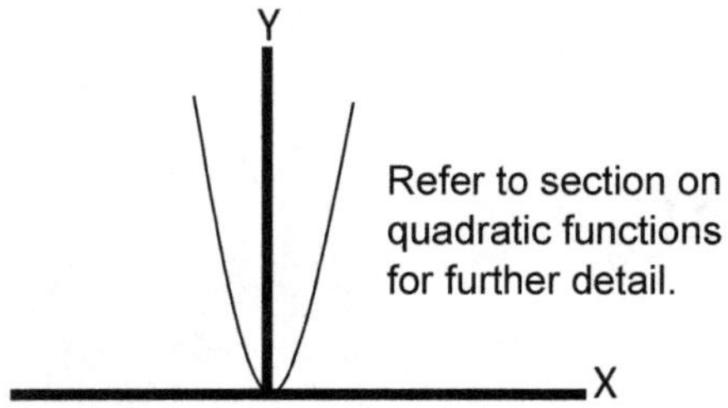

<u>Reciprocal:</u> $f(x) = 1/x$

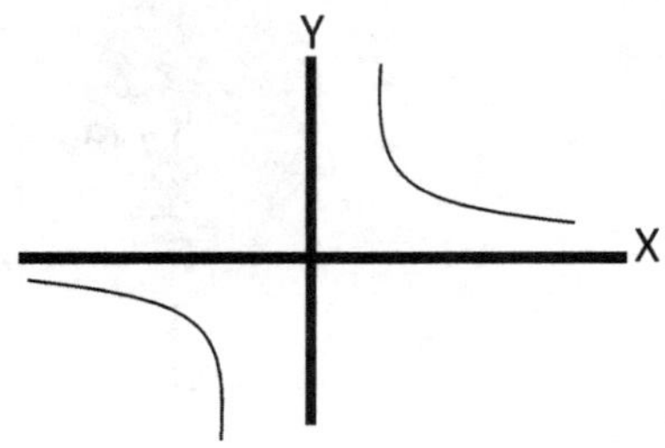

<u>Cube:</u> $f(x) = x^3$

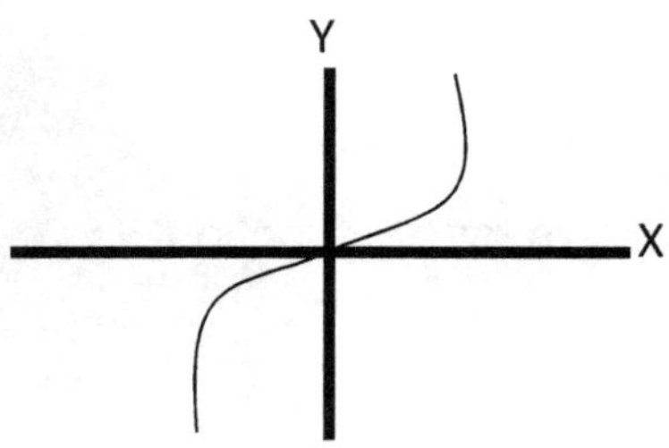

<u>Exponential:</u> $f(x) = e^x$

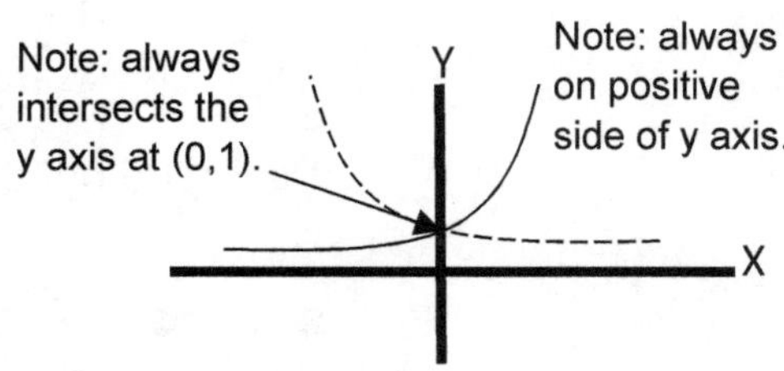

<u>Square root:</u> $f(x) = \sqrt{x}$

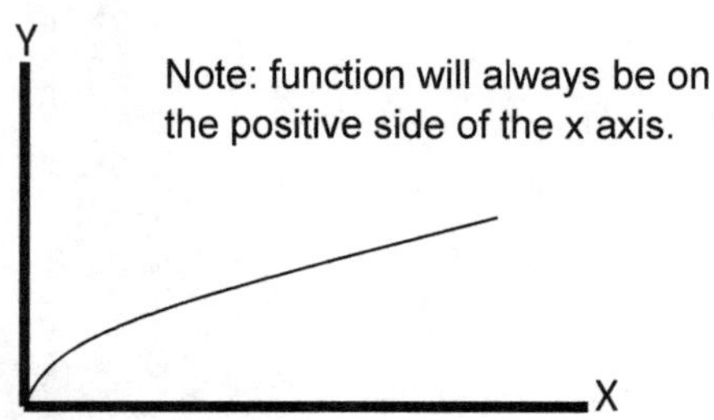

<u>Logarithmic:</u> $f(x) = \ln(x)$

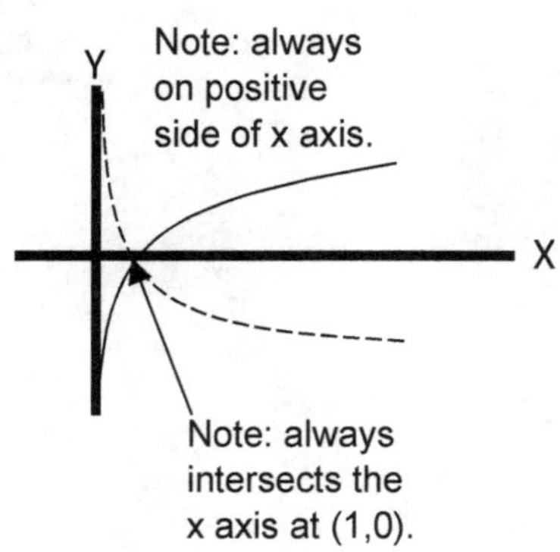

CONCLUSION AND REFERENCES

The information presented herein is aimed at providing the tools and knowledge to think critically and to reason quantitatively in an attempt to successfully pass the Fundamentals of Surveying (FS) examination.

Critical thinking- is the use and analysis of presented facts to produce a result or decision. It is to judge based on evidence. It is to analyze, contemplate, conceptualize, and evaluate a question and the facts/evidence it provides to make a decision and answer. The questions presented on the test will have facts provided to you. It will be your task to use these facts to derive the answer. The facts will be clues guiding you down a path, so use your knowledge of the material outlined herein to go further down the path and to find the answer.

Quantitative reasoning- is the use of mathematical skills to interpret practical information. It is the use of the Pythagorean Theorem of a right triangle to conclude the change in elevation between points during trigonometric leveling. It is the use of algebra and ratios to determine the location of a lost quarter section corner when single proportioning. Overall, survey mathematics is relative rudimentary and requires only a fundamental understanding of basic mathematic such as algebra, trigonometry, geometry, etc. It will be your task to find and apply the appropriate mathematic application to solve the problems.

This study manual is aimed to assist in understanding land surveying fundamentals and to progress one's career and path to licensure. The concepts of the following topics are explored: public lands survey system (i.e., section layout, township layout, quadrangle layout, single and double proportioning, closing corners, and meander lines), easements, title, deeds, descriptions, evidence, simultaneous conveyances, records research, rules of construction, order of operations, quadratic equations, triangles, degree to decimal degree conversions, bearing to azimuth conversions, magnetic declination, horizontal and vertical curves, area calculations, intersection of lines/slope, photogrammetry, statistics and probability, field measurements and corrections, level loops, geodesy, state plane and grid/ground conversions, and many others.

The material compiled herein is based on experience and land surveying education gained thoughout my career as a Professional Land Surveyor. Additionally, I applied common concepts and ideas from essential land surveying references including:

Brinker, Russell C., and Roy Minnick, eds. *The Surveying Handbook*. 2nd ed. New York, New York: Chapman & Hall, 1995.

Double Proportion. YouTube, 2012. https://www.youtube.com/watch?reload=9&v=hOrFFChbBfc.

FS Reference Handbook 9.0 Version for Computer-Based Testing, National Council of Examiners for Engineering and Surveying (NCEES), 2021.

Garner, Bryan A., and Henry Campbell Black. In *Black's Law Dictionary*. St. Paul, MN: Thomson Reuters, 2019.

"Global Positioning System." Wikipedia, August 23, 2021. https://en.wikipedia.org/wiki/Global_Positioning_System.

Hermansen, Knud. "SVT221: Boundary Law." *Surveying 221*. Lecture presented at the University of Maine Orono, January 2006.

"List of Principal and Guide Meridians and Base Lines of the United States." Wikipedia, February 25, 2021. https://en.wikipedia.org/wiki/List_of_principal_and_guide_meridians_and_base_lines_of_the_United_States.

Manual of surveying Instructions: For the survey of the public lands of the United States § (2009).

Public Land Survey System. Wikipedia, August 15, 2021. https://en.wikipedia.org/wiki/Public_Land_Survey_System.

Robillard, Walter G., Donald A. Wilson, and Curtis M. Brown. *Browns Boundary Control and Legal Principles*. 5th ed. Hoboken, New Jersey: John Wiley & Sons, Inc., 2003.

Sickle, Jan Van. *1001 Solved Surveying Fundamentals Problems*. 2nd ed. Belmont, California: Professional Publications, 2001.

"Types of Experimental Errors." Physics.NMSU.edu, June 21, 2021 http://www.physics.nmsu.edu/research/lab110g/html/ERRORS.html#:~:text=Errors%20are%20normally%20classified%20in,%2C%20random%20errors%2C%20and%20blunders.&text=Systematic%20errors%20are%20due%20to,high%20or%20consistently%20too%20low.

I strongly recommend acquiring material(s) for your own surveying library for routine use as well as studying purposes.

CPSIA information can be obtained
at www.ICGtesting.com
Printed in the USA
LVHW061507210223
740052LV00013B/931

9 780990 438366